Formation and Evolution of Earth's Crust

Edited by Mualla Cengiz and Savaş Karabulut

Published in London, United Kingdom

Formation and Evolution of Earth's Crust
http://dx.doi.org/10.5772/intechopen.107566
Edited by Mualla Cengiz and Savaş Karabulut

Contributors
Anderson Santos, André Ferrari, Estefan Fonseca, Hanan Zwair, Hieu Cong Duong, Kenji Motoki, Leonardo Barão, Linh Thuy Thi Hoang, Mohamed Ben Chelbi, Monica Da Costa Pereira Lavalle Heilbron, Nguyen Do Duc, Niem Van Nguyen, Oleksii Vatslav Bartashchuk, Peter Szatmari, Susanna Sichel, Tan Trong Bui, Thomas Campos, Tien Cong Dinh, Valeriy V. Snakin

First published in London, United Kingdom, 2024 by IntechOpen
IntechOpen is the global imprint of INTECHOPEN LIMITED, registered in England and Wales, registration number: 11086078, 167-169 Great Portland Street, London, W1W 5PF, United Kingdom

British Library Cataloguing-in-Publication Data
A catalogue record for this book is available from the British Library

Additional hard and PDF copies can be obtained from orders@intechopen.com

Formation and Evolution of Earth's Crust
Edited by Mualla Cengiz and Savaş Karabulut
p. cm.
Print ISBN 978-1-83769-186-9
Online ISBN 978-1-83769-187-6
eBook (PDF) ISBN 978-1-83769-188-3

For EU product safety concerns:
IN TECH d.o.o., Prolaz Marije Krucifikse Kozulić 3, 51000 Rijeka, Croatia,
info@intechopen.com or visit our website at intechopen.com.

Meet the editors

Dr. Mualla Cengiz is a professor in the Department of Geophysical Engineering, Istanbul University–Cerrahpaşa, Turkey. She researches paleomagnetism applied to the tectonic assemblages of Anatolia and its surrounding area. She lectures on gravimetry and magnetism of the Earth, rocks, and the environment. She has authored many international publications and collaborated on several national and international projects. She is actively researching geophysical applications in caves, salt resources, and seawater intrusion.

Dr. Savas Karabulut was an assistant professor in the Department of Seismology, Istanbul University, Turkey, from 2002 to 2016. He has been a lecturer in the Department of Civil Engineering, Gebze Technical University, Turkey, since 2021. His research interests are engineering seismology, soil dynamics, seawater intrusion, paleomagnetism, soil engineering problems, statistical hazard analysis, and cave/sinkhole investigation using geophysical methods. He has collaborated on many projects about earthquake mitigation, seismic hazards, urban transformation, and earthquake disasters. He has authored more than 100 scientific publications, books, and conference papers.

Contents

Preface **IX**

Section 1
Geophysical and Geological Applications to the Structure of the Earth's Crust **1**

Chapter 1 **3**
Interpretation of High-Resolution Remote Sensing Data for Plio-Pleistocene Tectonic Structures Studies in the Trindade Island Volcanic Building, South Atlantic Ocean, Brazil
by Kenji Motoki, Thomas Campos, Anderson Santos, Monica Heilbron, Leonardo Barão, Susanna Sichel, André Ferrari, Estefan Fonseca and Peter Szatmari

Chapter 2 **27**
Formation and Evolution of the Sarmatia Earth's Crust (East European Craton): Evidence from the Dnipro-Donets Paleorift
by Oleksii Bartashchuk

Chapter 3 **55**
New Insights into Geodynamic Evolution of the South-Eastern Termination of the Tunisian Atlas during Early Cretaceous Period from Surface and Subsurface Data: Hydrogeological Implications
by Mohamed Ben Chelbi

Section 2
Geochemistry and Mineralization of the Earth's Crust **89**

Chapter 4 **91**
Characteristic of Specialized Geochemistry of the Ngoc Tu Granitoid Block KonTum, Vietnam
by Nguyen Duc Do, Niem Van Nguyen, Hieu Cong Duong, Tan Trong Bui, Linh Thuy Thi Hoang and Tien Cong Dinh

Section 3
Plate Tectonics and the Evolution of Biodiversity **107**

Chapter 5 **109**
Perspective Chapter: Plate Tectonics and Global Dynamics of Biodiversity
by Valeriy V. Snakin

Chapter 6 **123**
Perspective Chapter: Ostracoda
by Hanan Zwair

Preface

The determination of discontinuities in the crust–mantle boundary and the lithosphere, the study of active and passive continental margins, and the mapping of divergent, convergent, and transform fault boundaries from the past to the present are among the most popular fields of geoscientific study in recent years. Due to the need for energy and ore deposits, it is important to determine the effects of mineralization on the Earth's crust, which can be done through the application of geosciences. Geophysical, geological, and geochemical studies now occupy an important place in determining heat flows and geothermal resources as well as the thermodynamic processes in the crust. This book includes studies on the evolution of oceanic and continental crust over different geologic time periods.

This book presents valuable results from geoscientific applications related to plate tectonics. Section 1, "Geophysical and Geological Applications to the Structure of the Earth's Crust", includes three chapters. Chapter 1, "Interpretation of High-Resolution Remote Sensing Data for Plio-Pleistocene Tectonic Structures Studies in the Trindade Island Volcanic Building, South Atlantic Ocean, Brazil" by Kenji Motoki et al., discusses two distinct events due to paleostresses according to remote sensing data correlated with the opening of the South Atlantic Ocean. Chapter 2, "Formation and Evolution of the Sarmatia Earth's Crust (East European Craton): Evidence from the Dnipro-Donets Paleorift" by Oleksii Bartashchuk, shows the stress field of Sarmatia after structural and kinematic analysis. Chapter 3, "New Insights into Geodynamic Evolution of the South-Eastern Termination of the Tunisian Atlas during Early Cretaceous Period from Surface and Subsurface Data: Hydrogeological Implications" by Mohamed Ben Chelbi, presents seismic and gravimetric data with the combination of the geodynamic evolution of the Menzel Habib Plain during the Early Cretaceous.

Section 2, "Geochemistry and Mineralization of the Earth's Crust", includes Chapter 4, "Characteristic of Specialized Geochemistry of the Ngoc Tu Granitoid Block KonTum, Vietnam" by Nguyen Duc Do et al., which highlights the oxidation-reduction state of granitoids and draws attention to the movements of molybdenum from magmatic solution to ore solution.

Section 3, "Plate Tectonics and the Evolution of Biodiversity", includes two chapters. Chapter 5, "Perspective Chapter: Plate Tectonics and Global Dynamics of Biodiversity" by Valeriy Snakin, concludes that the extinction of marine species occurred during plate merging and the increase in diversity occurred during plate separation. Finally, Chapter 6, "Perspective Chapter: Ostracoda" by Hanan Zwair, interprets the morphology of ostracods based on their adaptations to different environments.

Dr. Mualla Cengiz
Professor,
Department of Geophysical Engineering,
Istanbul University-Cerrahpaşa,
Istanbul, Turkey

Dr. Savaş Karabulut
Associate Professor,
Department of Civil Engineering,
Gebze Technical University,
Kocaeli, Turkey

Section 1

Geophysical and Geological Applications to the Structure of the Earth's Crust

Chapter 1

Interpretation of High-Resolution Remote Sensing Data for Plio-Pleistocene Tectonic Structures Studies in the Trindade Island Volcanic Building, South Atlantic Ocean, Brazil

Kenji Motoki, Thomas Campos, Anderson Santos, Monica Heilbron, Leonardo Barão, Susanna Sichel, André Ferrari, Estefan Fonseca and Peter Szatmari

Abstract

Trindade Island (20°31′0' S and 29°19′0 W) is a large alkaline volcanic edifice and is the only emerging part of the Vitória-Trindade ridge, 1140 km from the Brazilian offshore, whose summit is around 600 meters high. The lineament orientation pattern of Plio-Pleistocene dykes and fracture swarms identified by high-resolution remote sensing allowed identifying the paleo-stresses that acted on Trindade Island. Our research suggests that Trindade Island suffered two distinct tectonic events between 3.6 Ma and 0.25 Ma: the first event is related to the island formation and shows a NW-SE (compressional) and NE-SW (extensional) stress orientation, while the second event shows different stress, namely ENE-WSW (compressional) and NNW-SSE (extensional). These events may also be associated with the opening of the South Atlantic Ocean, during movement and torsion of the South American Plate in the NE-SW direction, evidencing compression toward ENE or due to the influence of the Trindade mantle plume on the South American Plate which generated intraplate stress.

Keywords: remote sensing, stress field, Plio-Pleistocene volcanism, Trindade Island, South Atlantic Ocean, Brazil

1. Introduction

The theoretical concept of plate tectonics, in which the lithosphere is divided into a mosaic of plates, which move and sink into the weaker ductile asthenosphere, is that tectonic plates are rigid and deformations only occur at convergent, divergent, and transform boundaries and that the driving force of plate drift is due to their subduction on convergent boundaries [1–6], but in recent years, has discovered the existence of intraplate deformation in Atlantic oceanic crust, associated to regional efforts due to mantle plume influence [7–17].

The South American Plate occupies a large area between the Caribbean Sea, the Peru-Chile Trench, Nazca Plate, Cocos Plate, and the Caribbean, and Atlantic Plates. It represents one of the regions with strong tectonic deformation in the Southern Hemisphere and has a peculiar tectonic feature (e.g., Andes Cordillera). This tectonic plate shows activities such as earthquakes and volcanic eruptions, which are due to the eastward drift of the Nazca Plate and westward expansion of the mid-Atlantic ridge that imposes intra- and inter-plate deformation and different drifts between them. This is a consequence of the long duration of Mid-Atlantic Ridge expansion, provoked not only by mantle plumes but also by whole-mantle convection, with ridge up-welling acting as a counterweight to crustal subduction [9, 18].

This jigsaw puzzle is responsible for several morphological features observed onshore and offshore worldwide: for example, Himalayas, Saint Andreas Fault, Mesoceanic Ridges, Alps, Mariana's Trench, and so on. On the Atlantic Ocean side, after Gondwana breakup, several transform faults structured its basement and supported the almost orthogonal opening of its counterparts [19]. In southeast Brazil, during Atlantic Ocean overture, several fracture zones were formed and extended continental inland forming preferential zones for magma ascending [20]. These structures supported magma ascending and generation and are associated with a complex involvement with an orogenic process starting in the Pangea fragmentation, going through the Brasiliano Orogeny and Gondwana breakup [21]. These authors also claimed for a shallow mantle plume in a convective system responsible for ascending hybrid metasomatized melt homogenized prior to eruption from a multivariate mantle reservoir, with elemental chemical imprint from ancient subducted oceanic slabs, and delaminated SLCM [22]. In this context, the Vitória-Trindade Ridge (**Figure 1**) has a controversial evolutionary history dividing authors between plate tectonics generation and its fracture zones [20, 24, 25] and a combined shallow mantle plume genesis [19].

The Atlantic Ocean is one of the least studied ones on the planet, with several volcanic edifices lacking geological and scientific studies. The Brazilian conjugated Margin has been studied since the past century, mostly in the platform area due to the oil & gas industry. As it advances ocean ward, indirect scientific studies (potential methods; seismic and other geophysics methods) have been conducted in the last 70 years. Concerning the geological studies of islands and seamounts, the first study was conducted by Almeida in the 1950s, and not before the 1970s, the third research has been conducted in the islands. Most of the knowledge we do have now related to the Meso-Cenozoic Brazilian offshore volcanics comes from the past 10 years of our research team, publishing over 15 papers. Most of them concerning the genesis and evolution of the biggest volcanic ridge in the Brazilian offshore area: the Vitória-Trindade Ridge (VTR), composed of more than 30 seamounts, banks, guyots and one island and one archipelago Trindade Island and Martin Vaz Archipelago (**Figure 1**).

In line with this research field, the objective of this chapter is to discuss the tectonic forces that control the magmatism on the Trindade Island, apparently associated with the Trindade Plume influence [7, 17], from the Interpretation of high-resolution remote sensing data for Plio-Pleistocene tectonic structures, like dykes and fault orientations. Dykes and faults orientation are indicators of tectonics, since they are positioned at 90° from the least stress tensor - σ_3 [26–29]. There is an agreement

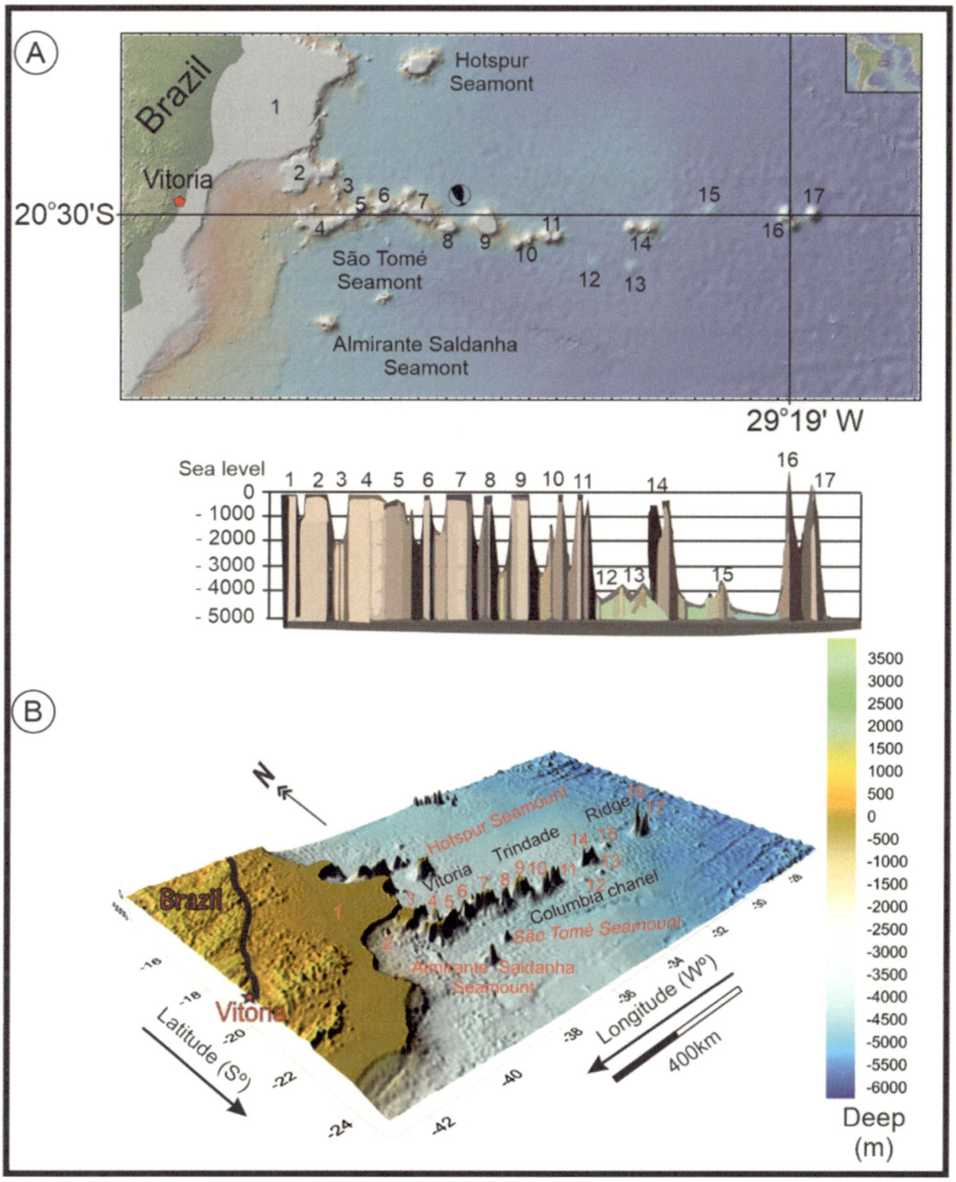

Figure 1.
(A) Trindade Island location, South Atlantic Ocean, Brazil (regional bathymetric map of the Brazilian southeastern continental margin, modified from the ridge multibeam synthesis (GeoMapApp (www.geomapapp.org) [23], (B) 3D color TDM view over Vitória-Trindade ridge (predicted bathymetry based on the TOPEX/Poseidon ver. 15.1, modified from [23]. 1– Abrolhos volcanic complex, 2– Vitória seamount, 3– Besnard Bank, 4– Champlain seamount, 5– Congress seamount, 3 nt, 6 – Montague, 7– Jaseur seamount, 8– Columbia seamount, 9– Davis Bank, 10– Asmus Bank, 11– Dogaressa Bank, 12– Gilberto Amado Hill, 13– Motoki Hill, 14– Columbia bank, 15– Palma seamount, 16– Trindade Island, 17– Martin Vaz archipelago. Black beachball: Nodal-plane solution from 6.5 earthquake of march/1955: 19.9°S; 36.7°W (based 87).

between σ_1 and σ_2, "calculated through both the orientation of dykes and faults" movement, at the same efforts field [30]. This relationship is also valid for volcanic areas and for intrusive bodies, and not only for purely extensive structures but also for transtensive ones [31].

2. Geological context

After the first work made by Almeida [32] in the Trindade Island, publishing the first geological map for Trindade Island (**Figure 2**), isolated publications about the Vitória-Trindade Ridge (VTR) raised a possible relationship between a mantle plume and oceanic lithosphere interaction in the South American Plate [37–41]. The VTR is an alignment composed of guyots, banks, seamounts, and islands, where Trindade Island and Martin Vaz Archipelago are inserted in the easternmost part of it. Several alkaline massifs inland Brazil territory are linked to the offshore extension, roughly parallel to the ca. 20°S, through a seamount ridge toward the islands 1200 km away from the Brazilian coast (**Figure 1**).

The seamounts and islands are linked to the Trindade Plume and they are from west to east (**Figure 1**): Abrolhos Magmatic Province (1), Besnard Seamount (2), Vitória Seamount (3), Hotspur Seamount, Congress Bank (4), Montague Seamount

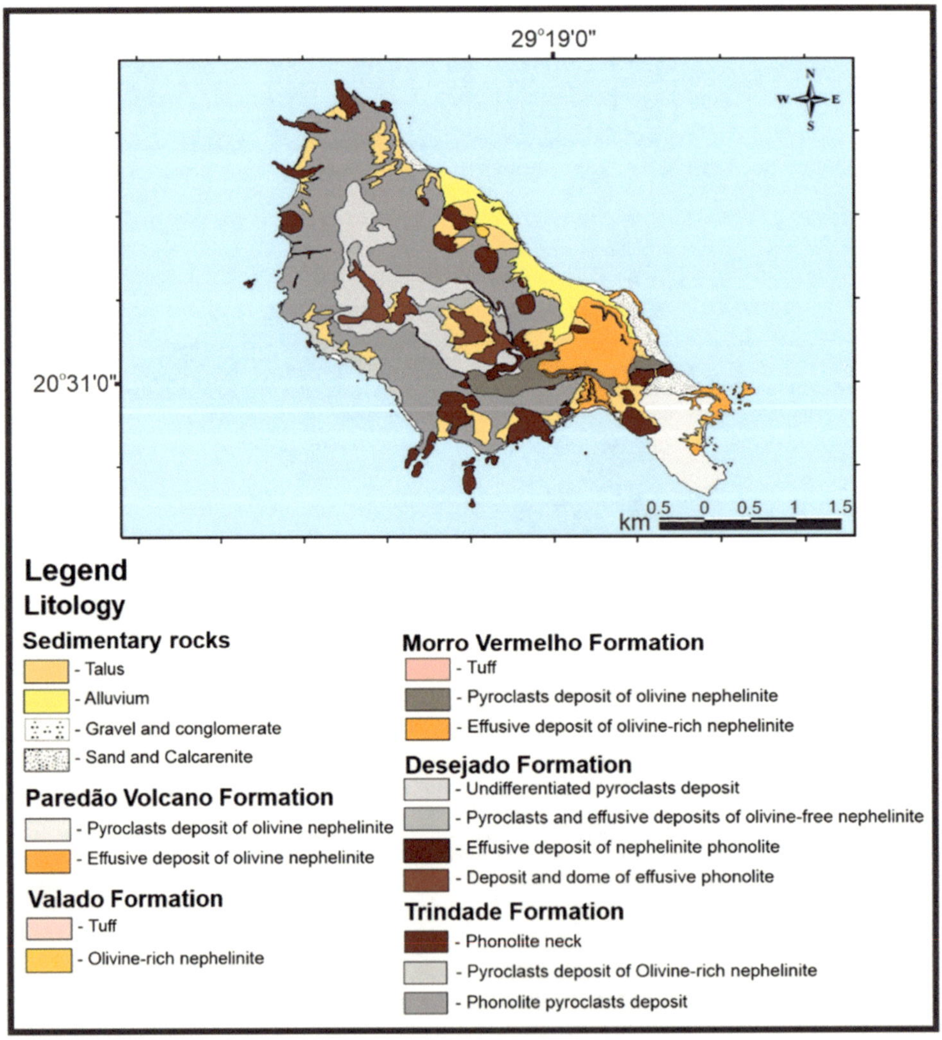

Figure 2.
Simplified geological map from the Trindade Island, Atlantic Ocean, Brazil. Based on [33–36].

(5), Jaseur Seamount (6), Colúmbia Bank (7), Davis Bank (8), Asmus Bank (8A), Dogaressa Bank (9), Colúmbia Seamount (10), Palma Seamount (11), and Trindade-Martin Vaz Archipelago. Motoki Hill (12) is at south Colúmbia Seamount (10), which includes Asmus Bank, Palma Seamount, and Motoki Hill [19, 42–48]. There are controversies about this magmatism, some authors related this magmatism to the Atlantic Ocean break-up in the Lower Jurassic [33, 37, 39, 49], while other researchers (e.g., [19–21, 25, 43, 44, 50–56]) concluded that the alkaline rocks are associated with the Trindade mantle plume. Some of these latter authors proposed that the original composition of Trindade plume magmas was enriched by oceanic lithospheric mantle fragments and detached subcontinental lithospheric blocks, left at shallow levels during the Gondwana breakup, and which underwent thermal remobilization by the Trindade plume.

Moreover, mantle metasomatism has been invoked to explain trace element enrichment and the highly alkaline nature of the volcanic rocks (e.g., [57–62]). Niu et al. [63] described the pre-metasomatic material source being the primitive mantle or a depleted reservoir left behind after extraction of the Atlantic Ocean crust. Ocean Island Basalt (OIB) is usually enriched in incompatible trace elements and this signature involves the need of a previous or simultaneous volatile-rich event melt infiltration in the mantle (F-; Br-; Cl-; H_2O; CO_2), enriching the melts in incompatible trace elements [64].

Trindade Island (TI) is situated in the South Atlantic Ocean at coordinates 20°31′0"S and 29°19′0"W, approximately 1140 km away from the Brazilian coast (see **Figures 1** and **2**). This island represents the exposed portion of a large alkaline volcanic structure, with its base submerged at a depth *circa* 5000 m. The island spans an area of 10.2 km^2 and rises approximately 625 m above the current sea level, featuring three distinct peaks. It is closely associated with the aseismic Vitória-Trindade Ridge, a significant geographic feature in the South Atlantic Ocean, which exhibits an east-west direction (see **Figure 1**).

The Vitória-Trindade Ridge (VTR) was formed from the passage of the South American Plate over the Trindade Plume (TP) in the Cenozoic [19, 24, 33–62, 64–75]. To Brazil, the Island's geology is an important preserved record of an undersaturated alkaline Plio-Pleistocenic volcanism. These geological features include volcanic relief such as phonolitic necks and dykes (e.g., nephelinite; sannaite; phonolite) and sub-horizontal layers of pyroclastic rocks and lava. These rocks were classified into five lithostratigraphic units [32]. These units represent the five distinct volcanic episodes associated with the island's formation and evolution.

The first volcanic event gave rise to the Trindade Formation (TF) composed of pyroclastic and subvolcanic rocks of diverse composition (e.g., basanites, lamprophyre, and phonolite). The four subsequent Formations (Desejado – DF; Morro Vermelho MVF; Valado – VF; and Paredão Volcano Formation – PVF) correspond to diversified pyroclastic and volcanic lithology (e.g., nephelinite and phonolite) (**Figure 2**) [25, 33–36].

$^{40}Ar/^{39}Ar$, $^{40}K/^{40}Ar$, and $^{14}7Sm/^{143}Nd$ dating [25, 67, 68] indicate aerial volcanism starting at 3.6 Ma and continuing up to 0.25 Ma, thus being one of the youngest volcanic episodes in the Brazilian geological record together with Martin Vaz Archipelago [52, 53, 69].

The VTR, as depicted in **Figure 1**, is believed to be linked to the reactivation of the Vitória-Trindade Fracture Zone (VTFZ) in conjunction with the TP [52–55, 69]. This ridge serves as a volcanic trail created by the TP on the South American plate. The VTFZ played a role as a conduit for the transportation of magma derived from the enriched mantle during this process [7, 17, 24, 41, 45, 50, 52–54, 63, 66, 69–72]. The VTR shows a general direction of E-W between 20° and 21°S, showing a geomorphological structure described by the alignment of banks and seamounts [52]. This ridge extends for about 1100 km, from the Besnard Bank and the Vitória Seamount south of Abrolhos Volcanic Complex (**Figure 1**). These submarine banks, seamounts, and guyots, with some of them presenting flat-topped shape, for example, Davis Bank located less than −100 m from actual sea level, rising from - 5000 m from the continental slope and foot ridge [7, 17, 21, 50, 70, 71]. The ridge is wider toward the west, due to the combination of the banks and seamounts by the algae and bryozoan carbonate reefs development [11, 73, 74].

3. Methodology

Remote sensing and digital image processing were used on the World View-2 satellite, which shows a multispectral resolution of 1.86 m and a Panchromatic resolution of 0.46 m, featuring 9 bands ranging from 400 to 1040 nm (**Table 1**). Those images are processed using the software ArcGIS and QGIS. The structural and volcanic features were established based on [77].

The scene dates from 2015/05/08 and was submitted to atmospheric correction and Pansharpen through ERDAS Imagine and ER Mapper software. The RGB 8–6–5 + PAN composition was demonstrated to be the best false color composition without the use of mathematical devices between bands. This composition enhances the dyke contacts in relation to their wall rocks and between the rocks (**Figure 3**; **Table 2**).

The 3D bathymetry model of the ocean floor was used in the Satellite data from the TOPEX/Poseidon database by the Institute of Oceanography of the University of California (UCSD–SIO), with "Global Topography 1 minute resolution v. 19.1" and

RGB 8–6–5	Near Infrared 2 (366–1040 nm)	Red Edge (705–745 nm)	Red (630–690 nm)
Pyroclastic rock (Trindade Formation)	Low	Low	Low
Volcanic rock (Trindade Formation)	Moderate	High	Moderate
Subvolcanic rock (Trindade Formation)	High	High	High
Pyroclastic rock (Desejado Formation)	Low	Moderate	Low
Pyroclastic rock (Morro Vermelho Formation)	Low	Moderate	Moderate
Pyroclastic rock (Paredão Volcano Formation)	Low	Moderate	Moderate
Pyroclastic and Olivine-Nephelinite Tuff (Valado Formation)	Moderate	Moderate	Moderate
Beaches related to weathering	Low	High	High
Dune Beaches	Very High	Very High	Very High
Vegetation	High	High	High

Table 1.
Spectral characteristics of the main lithologies of Trindade Island in the RGB 8–6–5 composition, based on [76].

Figure 3.
Satellite imagery: False color RGB 8–6–5 + PAN hybrid composition from Trindade Island, South Atlantic Ocean.

	Dyke	Faults
Length (m)	5–665	4–283°
Geometric Mean	59	34
Standard Deviation	40.14	27.28
n	980	1188

Table 2.
Summary of dykes and faults identified by remote sensing imagery.

processed by software ERDAS IMAGINE, with the tool Rapid Atmospheric Correction. This data comes from the analysis of the artificial satellite orbits, which provides local gravity variations from each area scanned, this makes it possible to acquire data on a global scale with the exception of the northernmost part of the planet, where the satellite does not orbit. After correcting the data, the first free-air anomaly points are created and then a global map is generated. The combination of the free-air anomaly and known reference data (benchmarks) that were acquired through conventional ship bathymetry, allows the creation of a bathymetric map on a world scale or the predicted bathymetry map, with an apparent resolution of 900 m. Although these techniques do not have a detailed resolution, they make it possible to cover a large area and explore all kinds of features and morphologies around the globe. In the study, the data obtained from the TOPEX/Poseidon database were tied through a regular kriging mesh to interpolate the local data using the software ArcGIS (Spatial analyst tools: kriging).

After processing all the data, we utilized ArcGIS and QGIS software to analyze the images, delineating the structural framework and identifying the main volcanic features observed on Trindade Island.

4. Results and discussion

Based on the hybrid composition RGB 8–6–5 + PAN, it was possible to distinguish the main lithology that comprises the Island (**Figure 3**; **Table 1**), subdivided into five Formations [32, 36].

It should be noted that there are areas with high reflectance in all bands (**Figures 3** and **4**), which is due to the weathering of mafic minerals, which generates soils and sand riches in silica. Due to their bright color compared to the different geological structures around them, they can present a reflectance of up to 90% [76–78].

The rocks of the basal Trindade Formation (TF) are predominantly composed of phonolite and nephelinite of pyroclastic nature, extrusive and subvolcanic character [32, 36]. Extrusive phonolites usually have high reflectance in the Red Edge band (705–745 nm), exhibiting green colors ([78] and Reference therein). Subvolcanic phonolites have high reflectance in all bands, exhibiting a whitish color and, when altered show a reddish color, indicating high reflectance in Near Infrared 2 (860–1040 nm, see **Table 1**). Finally, pyroclastic rocks have low reflectance in all bands, exhibiting a bluish color (**Figure 4A**). The presence of whitish to reddish subvolcanic dykes cutting through the pyroclastic is also notable. Finally, the pyroclastic rocks exhibit a darker color, indicating low reflectance in all bands (**Figure 4B, C**).

Desejado Formation (DF) is composed of phonolite and nephelinite flows and various pyroclastic flows. The pyroclastic rocks have a similar spectral response to the pyroclastic rocks of the Trindade Formation and exhibit high reflectance in the Red

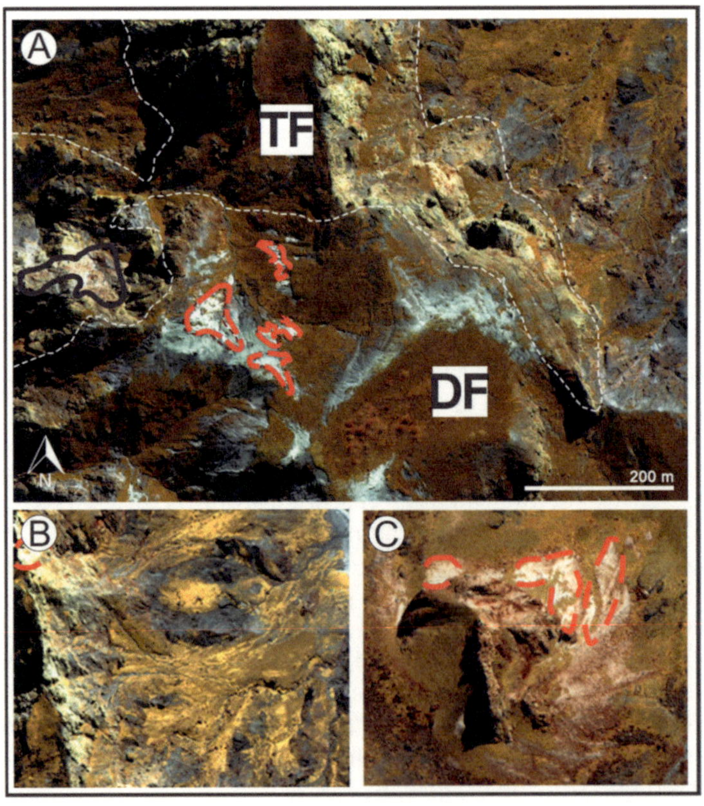

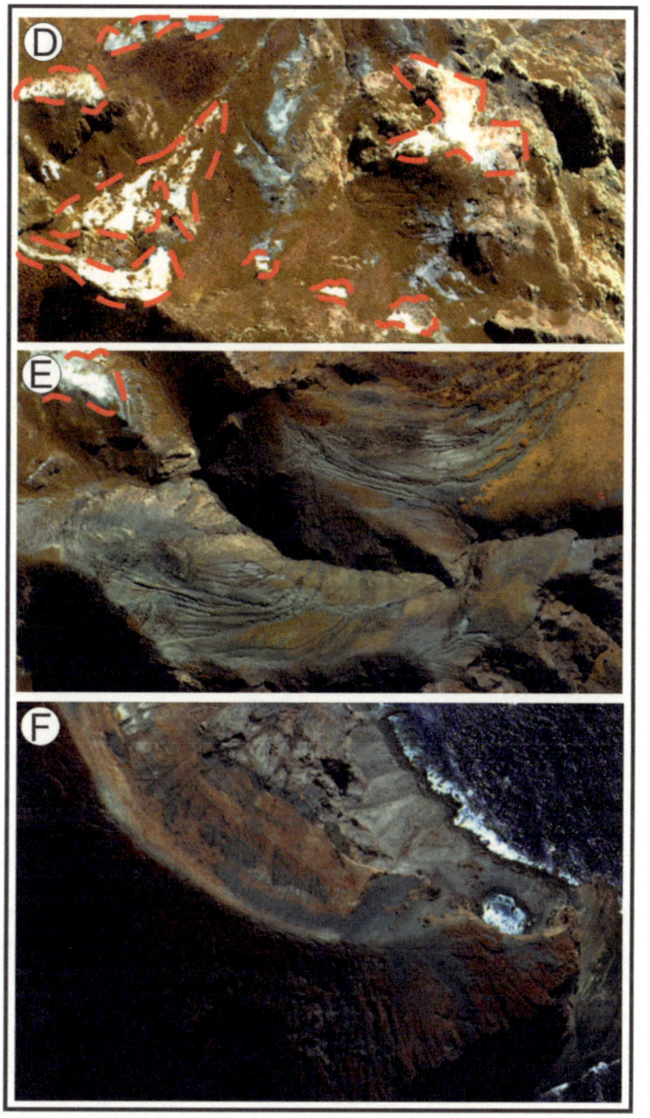

Figure 4.
(A) Contact between the subvolcanic rocks of the Trindade formation (TF) and the pyroclastic rocks of the Desejado formation (DF) is shown by white dashed line; red dashed line shows silica-rich residual soils and in black line delimited the weathered areas of the subvolcanic rocks from the TF; (B) contact between yellowish-green colored extrusive rocks and bluish colored pyroclastic rocks from the TF. Red to white colored subvolcanic dykes are visible, cutting through the pyroclastic rocks; (C) phonolite plug is white colored, showing alteration products in white-red color. The white areas correspond to weathered areas enriched in silica; (D) phonolite and nephelinite flows from Desejado formation show white color that corresponds to weathered areas enriched in silica (red line) and pyroclasts rocks showing bluish-green color, similar to the rocks of the Trindade formation; (E) pyroclastic rocks from Morro Vermelho formation show textures related to pyroclastic bedding and white colored area corresponds to weathered areas enriched in silica; (F) rocks from Paredão volcano formation show moderate reflectance in red and red edge and low reflectance in near-infrared 2. Red dashed line shows soils and sand riches in silica.

band (705–745 nm, see **Table 1**). The phonolite and nephelinite flows also resemble strew and subvolcanic rocks from the Trindade Formation. However, they have a higher reflectance and display intense white colors and a lower level, which correspond to alteration products of mafic minerals (**Figure 4D**).

Morro Vermelho Formation (MVF) is essentially composed of nephelinite and pyroclastic flows. The spectral response of both rocks is similar to the pyroclasts of the Trindade Complex and Desejado Formation (see **Table 1**). However, the reflectance in these rocks is lower, with a decrease in the Red band and an increase in the Red edge band response, showing textures related to the pyroclastic bedding (**Figure 4E**). The white colored area corresponds to weathered areas enriched in silica.

Valado Formation (VF) rocks are nephelinite tuffs and pyroclastic flows, outcrop in rare and little area of the Island (20°30′03.6″S and 29°19′15.6″W), being characterized by moderate reflectance in all bands, presenting a whitish color with red alteration products (see **Table 1**).

Paredão Volcano Formation (PVF) is composed of nephelinite tufts and pyroclastic, and presents a spectral response similar to the rocks from the Morro Vermelho Formation. It exhibits moderate reflectance in the Red band and moderate reflectance in the Red Edge band (see **Table 1**), and shows a greenish-to-bluish coloration. It is remarkable that the presence of altered areas, with moderate reflectance in the Near Infrared 2 band, exhibits a red color (**Figure 4F**).

The underwater portion of the Island was mapped using the RGB 3–2–1 composition, which is based on the Coastal Blue and Blue bands, presenting great penetrability in water bodies and allowing the underwater features visualization (**Figure 5**).

Our interpretations of satellite imagery allowed building a geological map of Trindade Island (**Figure 6**), with the lithology divided into five Formation. The phonolite domes and necks were distinguished from the other lithologies from Trindade Formation because these bodies are intensely fractured and show several dykes cutting them. The satellite-interpreted geological map of **Figure 6** is very similar to the field-interpreted geological map from **Figure 2** based on [32, 36].

Figure 5.
Trindade Island and its underwater surroundings, viewed through the RGB 3–2–1 composition satellite imagery.

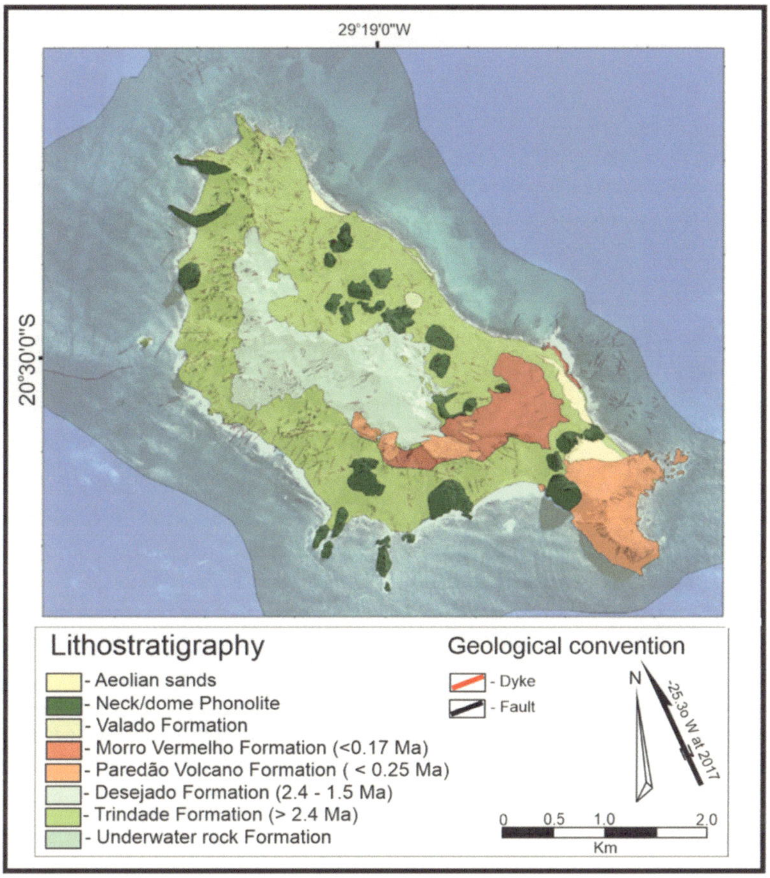

Figure 6.
Simplified geological map of Trindade Island based on satellite imagery.

4.1 Tectonic structure view by remote sensing imagery

The lineaments are roughly linear surface elements directly interpreted from remote imagery and can be related to structural tectonic movements. Dykes and faults are excellent tectonic indicators because they are positioned at 90° from the extension vector σ_3 [31, 79–83].

The lineaments identified by this work were divided into two categories: dykes (show thickness) and faults (no thickness). This distinction is important for a better understanding of the tectonic process throughout the Trindade Island genesis (**Figure 6**).

It identified 979 dykes and 1188 faults (see **Table 2**) that crop out in emerged and underwater rocks of the Trindade Island (**Figure 7**). **Figure 6** shows a simplified geological map based on satellite imagery, where can see all dykes and faults from TI and **Figure 8** shows the Rose diagram with the projections of respective directions.

Projection of the poles from lineament directions at Lower-hemisphere, equal-area stereo-plot [84], shows two general directions at N50°W and N75°E and the straight dihedral-angle method [84] defines the following compression axes σ_1 117/90 and σ_3 27/90 (**Figures 9** and **10**, **Table 3**). The stress ax direction related to the underwater dykes is located approximately 90° from the crop-out rocks, suggesting a compression and further stretching related to the Vitória-Trindade Fracture Zone [24].

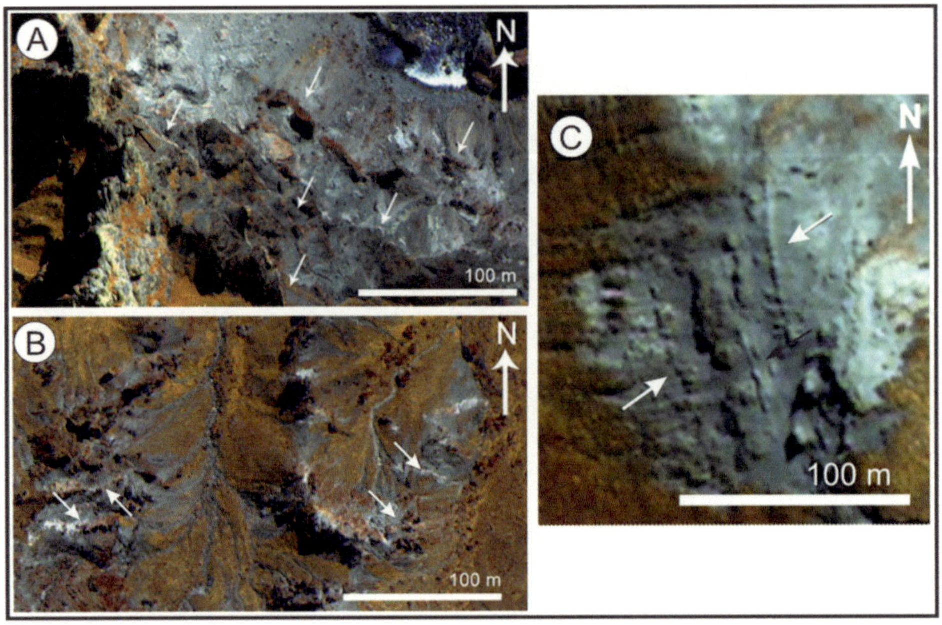

Figure 7.
(A) Dykes cutting the Trindade formation, with NW-SE trend; (B) dykes cutting Desejado formation, with ENE-WSW trend; (C) dykes cutting Desejado formation, with trend near NNW–SSE and presenting different spectral responses ***Figure 7.*** *(A) Dykes cutting the Trindade formation, with NW-SE trend; (B) dykes cutting Desejado formation, with ENE-WSW trend; (C) dykes cutting Desejado formation, with trend near NNW–SSE and presenting different spectral responses.*

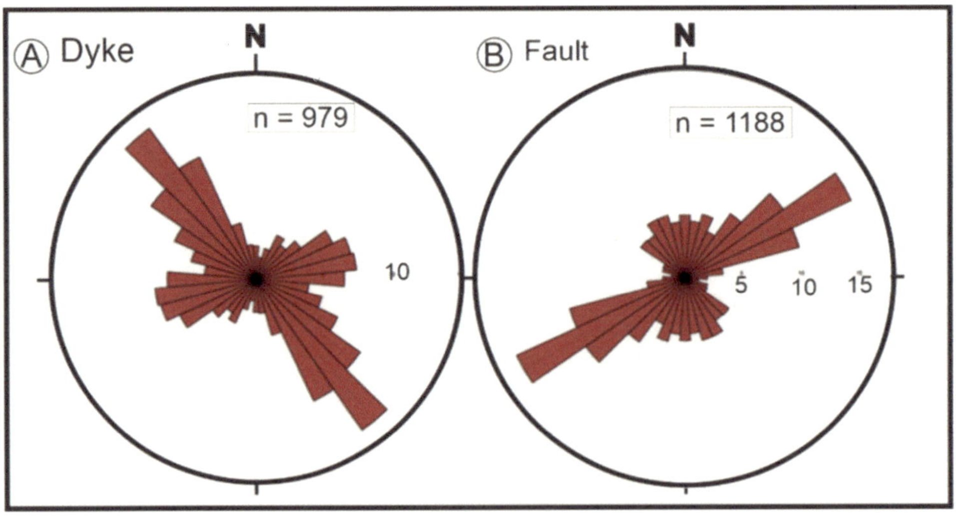

Figure 8.
Rose diagram projection of dykes (A) and faults (B) direction identified by satellite imagery from Trindade Island.

The faults show a predominantly NE-SW orientation, and more rarely NW-SE and NNE-SSE (**Figure 10A–E**; **Table 4**). The only earthquake recorded in the last 68 years along the Vitória-Trindade Fracture Zone is characterized by maximum compression

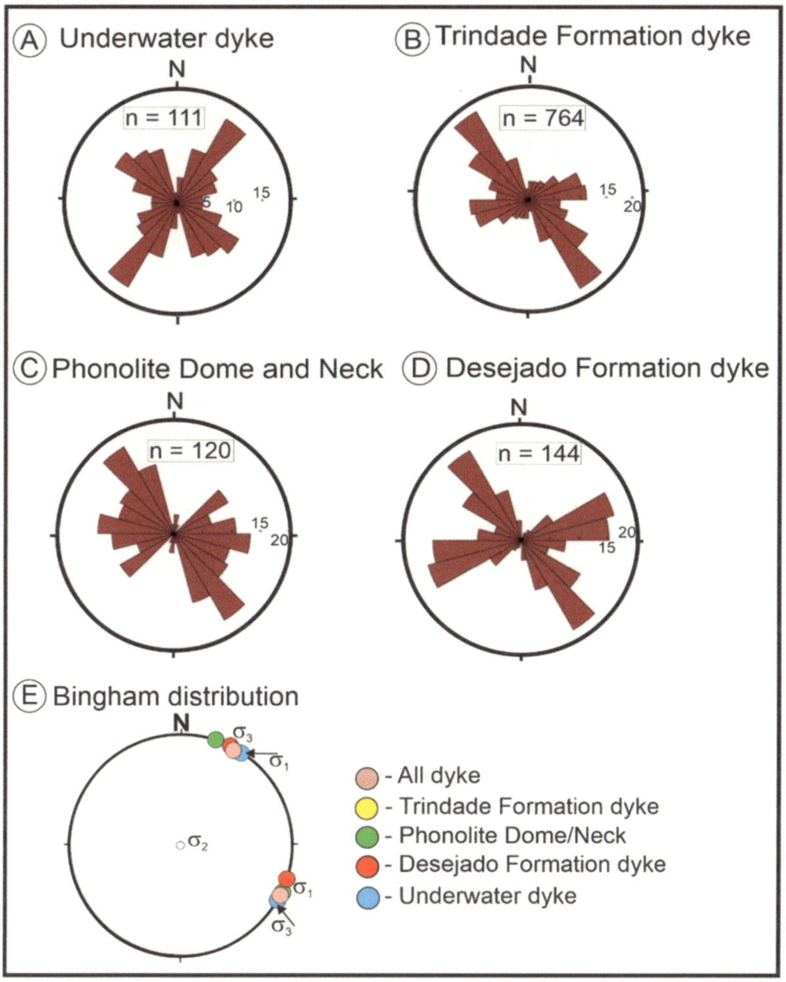

Figure 9.
Rose diagram with the directions of the dykes from the outcrop and underwater areas of Trindade Island identified from remote sensing imagery: (A) underwater dykes; (B) Trindade formation dykes; (C) phonolite dome and necks; (D) Desejado formation dykes; (E) Bingham distribution [66] with the dyke stress axes from Trindade Island.

in the ENE direction compatible with the formation of open old fractures identified on Trindade Island [20, 85]. Mendiguren & Richter [86] assert that the existence of intraplate stress in the middle South America Plate is due to thrust-faulting mechanisms for intraplate earthquakes that show a dominant horizontal deviatory compressional stress oriented in a NW-SE direction and this stress is due to forces connected with spreading center (mantle plumes) on the Mid Atlantic Ridge (**Figure 1**).

According to the seismic analysis conducted by the same authors, the epicenter of an earthquake (March/1955) with a magnitude of 6.4 was located slightly north (19.9°S, 36.7°W) of the Victoria-Trindade ridge (see **Figure 1**). The geological setting of this specific region suggested that the source mechanism of the earthquake might be attributed to stresses of topographic origin associated with the Victoria-Trindade ridge and Trindade mantle plume [86, 87]. The direction was similar to the one related to the formation of the Trindade Island which we defined as the first event,

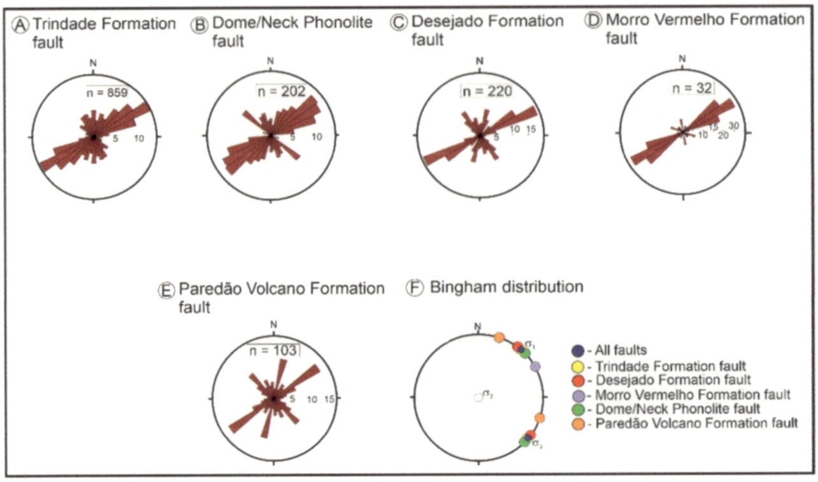

Figure 10.
Rose diagram with the fault direction from Trindade Island identified by remote sensing imagery: (A): Trindade formation; (B): Dome and neck of phonolite; (C): Desejado formation; (D): Morro Vermelho formation; (E) Paredão volcano formation; (F) Bingham distribution [45] of fault tension axes from Trindade Island.

	σ_1	σ_3
All dykes	117°	27°
Underwater dyke	33°	120°
Trindade Formation dyke	119°	29°
Desejado Formation dyke	108°	26°
Phonolite Dome and Neck	116°	17°

Table 3.
Summary of dyke directions identified by remote sensing imagery.

with NW-SE (compressional) and NE-SW (extensional) direction. Another event happened later, this one is characterized by ENE-WSW (compressional) and NNW–SSE (extensional) direction, we call this the second event.

The analysis of the fractures on Bingham distribution [84] shown in **Figure 10F** and **Table 4** suggests NE–SW direction for σ_1 (compression) and NW-SE for σ_3 (distension).

	σ_1	σ_3
All Faults	40°	130°
Trindade Formation fault	40°	130°
Desejado Formation fault	36°	126°
Morro Vermelho Formation fault	60°	130°
Paredão Volcano Formation Fault	18°	135°
Phonolite Dome and Neck	45°	108°

Table 4.
Summary of faults identified by remote sensing imagery.

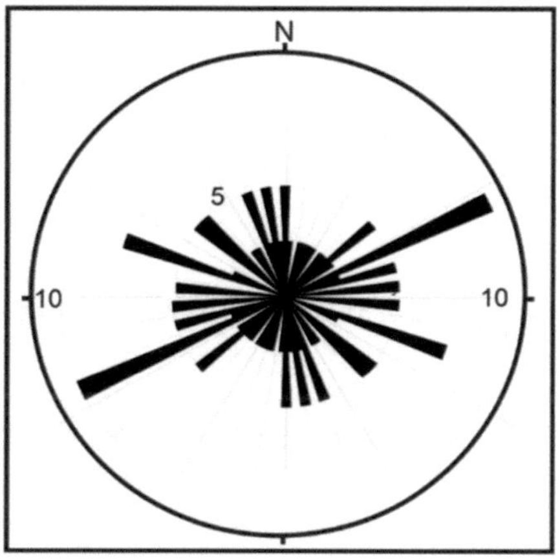

Figure 11.
Rose diagram with the conjugate faults directions from Trindade Island identified by remote sensing imagery.

All dyke directions are equivalent to those described by [20, 34, 36, 72, 85]. The NE-SW dykes from the Desejado Formation also suggest a NE-SW shortening explained by the faults. However, the rocks of the Desejado Formation are heavy vegetation cover, requiring field confirmation of our measurement interpretation.

Shortening axes (σ_1) inferred from fractures can only be used if these are open fracture, similar to dykes. The analysis of the few conjugated pairs from Trindade Island (n: 20) shows much dispersed data, with a possible E-W compression (**Figure 11**).

The dykes present a general direction of NW-SE, secondarily to ENE-WSW. When inserted in the Trindade Formation or Desejado formation, they replicate this general orientation. In the most recent unit (Desejado Formation), the occurrence of similar amounts of ENE-WSW dykes is notable.

Field data indicate a vertical dip, suggesting a simple shear regime with a vertical dip [20, 34, 36, 72, 85]. Assuming all dykes are perpendicular to the lowest regional extension vector σ_3 [81] and the paleo-stress analysis through the Bingham distribution [84] suggests shortening (σ_1) at N28W, similar to the values from [20, 72, 85]. For ENE-WSW dykes, the maximum shortening direction (σ_1) is N81E.

The Faults from Trindade Island present a general trend of ENE-WSW. Assuming a distensile character in a simple shear regime (e.g., T-joints of the Riedel fracture system), the direction of shortening (σ_1) at ENE-WSW and extension (σ_3) at NNW-SSE is inferred.

5. Conclusion

Assuming that the dykes are lodged in extensive fractures, which the fractures denote the same character, it was inferred that the paleo-tensions acting in the Trindade Island presented two distinct events throughout its evolution. The first event is related to the island genesis and is characterized by NW-SE compression and NE–SW extension, the second event is characterized by compression at ENE-WSW direction and extension at NNW–SSE direction. These old stresses are similar to the stress from the latest earthquake of magnitude 6.4 recorded in 1955 in the Vitória-Trindade Submarine Volcanic Ridge [86]. Our data reveal, therefore, that the Trindade Island suffered two distinct tectonic events between 3.6 and 0.25 Ma. Such events can be associated with the torsion that the South American plate suffered during the opening of the Atlantic Ocean or effects of the passage of the South American plate over the Trindade mantle plume, which causes spot stress field [88].

Acknowledgements

Fieldwork in the Trindade Island was only possible due to the cooperation with the Brazilian Navy and National Council for Scientific Research (CNPq) through the PRO-TRINDADE Program, which is responsible for the supervision and maintenance of scientific research in different areas of knowledge for this region of the Brazilian territory. The present work is linked to the Ministry of Science and Technology and Innovation and National Council for Scientific Research (MCT/CNPq project n ° 442805/2015-2), under the coordination of researchers from Federal University of Rio Grande do Norte. The Brazilian Navy maintains a military base at the Trindade Island Oceanographic Post (POIT), which included a scientific station of the Trindade Island for research and lodging of the researchers (ECIT).

Conflict of interest

The authors declare no conflict of interest.

Notes/thanks/other declarations

We thank the Brazilian Navy cooperation during the field work and operation, particularly the staff at the POIT, and support through the PRO-TRINDADE Program (Scientific Research Program on Trindade Island). We are also grateful to the LEPLAC Project for providing the seismic dataset in the deep-water region of the Espírito Santo Basin. We thank Digital Globe Foundation provided the imagery from WorldView-2 Satellite and the Hexagon Geospatial provided the Erdas Imagine and ER Mapper software.

DOI: http://dx.doi.org/10.5772/intechopen.113254

Author details

Kenji Motoki[1], Thomas Campos[2]*, Anderson Santos[3,4], Monica Heilbron[3,4], Leonardo Barão[5], Susanna Sichel[1], André Ferrari[1], Estefan Fonseca[1] and Peter Szatmari[6]

1 Department of Geology and Geophysics, Fluminense Federal University, Rio de Janeiro, Brazil

2 Department of Geology, Federal University of Rio Grande do Norte, Natal, Brazil

3 Department of Mineralogy and Igneous Petrology, Estate University of Rio de Janeiro Rio de Janeiro, Brazil

4 Tektos Group, GeoAtlântico Group and Geobiotec Group, Geoscience Department, Aveiro University, Portugal

5 Department of Geology, Federal University of Parana, Curitiba, Brazil

6 Leopoldo Américo Miguêz de Mello Research and Development Center (CENPES/PETROBRAS), Rio de Janeiro, Brazil

*Address all correspondence to: thomas.campos@ufrn.br

References

[1] McKenzie DP, Parker RL. The North Pacific: An example of tectonics on a sphere. Nature. 1967;**216**(5122):1276-1280. DOI: 10.1038/2161276a0

[2] Morgan WJ. Hotspot tracks and the early rifting of the Atlantic. Developments in Geotectonics. 1983;**19**:123-139. ISBN 9780444421982. DOI: 10.1016/B978-0-444-42198-2.50015-8

[3] Müller RD, Seton M, Motion P. Encyclopedia of Marine Geosciences. Dordrecht: Springer; 2014. pp. 1-10. DOI: 10.1007/978-94-007-6644-0_131-1

[4] Stern RJ. The evolution of plate tectonics. Philosophical Transactions of the Royal Society A: Mathematical, Physical and Engineering Sciences. 2018;**376**(2132):20170406. DOI: 10.1098/rsta.2017.0406

[5] Wessel P, Müller RD. Plate tectonics. Treatise on Geophysics. 2007;**6**:49-98. DOI: 10.1016/B978-0-444-53802-4-00111-1

[6] Wilson JT. A new class of faults and their bearing on continental drift. Nature. 1965;**207**:343-347. DOI: 10.1038/207343a0

[7] Conceição JCJ et al. Controle tectônico do magmatismo do complexo vulcânico de Abrolhos, Bacia do Espírito Santo. Congresso Brasileiro de Geologia. 1996;**39**:384-387

[8] Fernandez C et al. Shear zones as a result of intraplate tectonics in oceanic crust: The example of the basal complex of Fuerteventura (Canary Islands). Journal of Structural Geology. 1997;**19**(1):41-57. DOI: 10.1016/S0191-8141(96)00074-0

[9] Fleitout L, Dalloubeix C, Moriceau C. Small-wavelength geoid and topography anomalies in the South Atlantic Ocean: A clue to new hot-spot tracks and lithospheric deformation. Geophysical Research Letters. 1989;**16**(7):637-640. DOI: 10.1029/GL016i007p00637

[10] Gomes PO. O Projeto Leplac na margem continental nordeste: imagens sísmicas de flexura litosférica junto a zonas de fratura. In: 5th International Congress of the Brazilian Geophysical Society. São Paulo: European Association of Geoscientists & Engineers; 1997. DOI: 10.3997/2214-4609-pdb.299.9

[11] Gorini MA, Carvalho JC. Geologia da margem continental inferior brasileira e do fundo oceânico adjacente. In: Schobbenhaus C, Campos DA, Derze GR, Asmus HE, editors. Geologia do Brasil: Texto explicativo do Mapa Geológico do Brasil e da área oceânica adjacente incluindo depósitos minerais – Escala 1:2500000. Brasília: MME/DNPM; 1984. pp. 473-489

[12] Jordahl KA et al. Why there are no earthquakes on the Marquesas fracture zone. Journal of Geophysical Research: Solid Earth. 1995;**100**(B12):24431-24447. DOI: 10.1029/94JB02886

[13] Kruse SE et al. Evolution and strength of Pacific fracture zones. Journal of Geophysical Research: Solid Earth. 1996;**101**(B6):13731-13740. DOI: 10.1029/96JB00645

[14] Lowrie A, Smoot C, Batiza R. Are oceanic fracture zones locked and strong or weak?: New evidence for volcanic activity and weakness. Geology. 1986;**14**(3):242-245. DOI: 10.1130/0091-7613(1986)14<242:AOFZLA>2.0.CO;2

[15] Marinoni LB, Pasquarè G. Tectonic evolution of the emergent part of a volcanic ocean island:

Lanzarote, Canary Islands. Tectonophysics. 1994;**239**(1-4):111-137. DOI: 10.1016/0040-1951(94)90110-4

[16] Mello SLM, Dias MS. Magnetoestratigrafia da crosta oceânica entre as zonas de fratura de Ascensão e Bode Verde. Brazilian Journal of Geophysics. 2018;**14**(3):237-252. DOI: 10.22564/rbgf.v14i3.1222

[17] Szatmari P, Mohriak WU. Plate model of post-breakup tectono-magmatic activity in SE Brazil and the adjacent Atlantic. Simpósio Nacional de Estudos Tectônicos. 1995;**1995**(5):213-214

[18] Agius MR et al. A thin mantle transition zone beneath the equatorial mid-Atlantic ridge. Nature. 2021;**589**(7843):562-566. DOI: 10.1038/s41586-020-03139-x

[19] Santos AC, Hackspacher PC. Meso-Cenozoic Brazilian Offshore Magmatism: Geochemistry, Petrology and Tectonics. 1st ed. VL 1. London: Academic Press, Elsevier; 2021. 300 p. ISBN: 978-0-12-823988-9. DOI: 10.1016/C2019-0-05400-3

[20] Alves EC et al. Ocean fracture zones: Their evolution and impact on tectonic and magmatism of the south and southeast Brazilian continental margin. In: Meso-Cenozoic Brazilian Offshore Magmatism. London: Academic Press; 2022. pp. 47-94. DOI: 10.1016/C2019-0-05400-3

[21] Quaresma GDOA et al. Isotopic constraints on Davis bank, Vitória-Trindade ridge: A revised petrogenetic model. Journal of South American Earth Sciences. 2023;**122**:104099. DOI: 10.2139/ssrn.4059025

[22] Szameitat LSA et al. Geophysical evidence for lithospheric scale asymmetry and inherited mantle in the SE Brazilian-Angola and Newfoundland-Iberia rifted margins. Journal of South American Earth Sciences. 2023;**123**:104214. DOI: 10.1016/j.jsames.2023.104214

[23] Ryan WBFM et al. Global multi-resolution topography (GMRT) syntese data set. Geochemistry, Geophysics, Geosystems. 2009;**10**:QQ03014. DOI: 10.5069/G9BG2M6R

[24] Alves EDC et al. Zona de fratura de Vitória-Trindade no Oceano Atlântico sudeste e suas implicações tectônicas. Revista Brasileira de Geofisica. 2006;**24**:117-127. DOI: 10.1590/S0102-261X2006000100009

[25] Marques LS et al. Petrology, geochemistry and Sr–Nd isotopes of the Trindade and Martin Vaz volcanic rocks (southern Atlantic Ocean). Journal of Volcanology and Geothermal Research. 1999;**93**(3-4):191-216. DOI: 10.1016/S0377-0273(99)00111-0

[26] Caputo R. A comparison between joints and faults as brittle structures used for evaluating the stress field. Annales Tectonicae. 1991;**5**(1):74-84

[27] Hancock PL. Determining contemporary stress directions from neotectonic joint systems. Philosophical Transactions of the Royal Society of London Series A: Physical and Engineering Sciences. 1991;**337**(1645):29-40. DOI: 10.1098/rsta.1991.0104

[28] Mandl G. Rock Joints. 5th ed. Berlin: Springer; 2005. 222 p. ISBN-13 978-3540245537

[29] Ramsay JG, Huber MI, Lisle RJ. The Techniques of Modern Structural Geology: Folds and Fractures. Vol. 2. London: Academic Press, Elsevier; 1987. p. 500. ISBN-10:0-12-576922-9

[30] Geoffroy L, Bergerat F, Angelier J. Modification d'un champ de contrainte regional par un champ de contraintes magmatiques local: Exemple de l'ile de Skye (Ecosse) au Paleocene. Bulletin de la Société Géologique de France. 1993;**164**(4):541-552

[31] Bassett KN, Kleinspehn KL. Mid-cretaceous transtension in the Canadian cordillera: Evidence from the rocky ridge volcanics of the Skeena group. Tectonics. 1996;**15**(4):727-746. DOI: 10.1029/95TC03496

[32] Almeida FFM. Geologia e petrologia da Ilha da Trindade. Brazil: Serviço Gráfico do Instituto Brasileiro de Geografia e Estatística; 1961. Available from: https://repositorio.usp.br/directbitstream/e9854a6e-ad60-4192-828d-7ca6c5668765/FT-328.pdf

[33] Toyoda K et al. Dupal anomaly of Brazilian carbonatites: Geochemical correlations with hotspots in the South Atlantic and implications for the mantle source. Earth and Planetary Science Letters. 1994;**126**(4):315-331. DOI: 10.1016/0012-821X(94)90115-5

[34] Almeida, F.d. As ilhas oceânicas brasileiras e uma hipótese sobre a origem do Atlântico. Anais da Academia Brasileira de Ciências. 1965;**37**:141-145

[35] Pires GLC, Bongiolo EM. The nephelinitic–phonolitic volcanism of the Trindade Island (South Atlantic Ocean): Review of the stratigraphy, and inferences on the volcanic styles and sources of nephelinites. Journal of South American Earth Sciences. 2016;**72**:49-62. DOI: 10.1016/j.jsames.2016.07.008

[36] Patricio RL. Mapeamento da Ilha da Trindade, Atlântico Sul [thesis]. Brazil: Universidade Estadual Paulista Júlio de Mesquita Filho, Instituto de Geociências e Ciências Exatas; 2012. Available from: http://hdl.handle.net/11449/120425

[37] Gibson SA et al. The late cretaceous impact of the Trindade mantle plume: Evidence from large-volume, mafic, potassic magmatism in SE Brazil. Journal of Petrology. 1995;**36**(1):189-229. DOI: 10.1093/petrology/36.1.189

[38] Gibson SA et al. The limited extent of plume-lithosphere interactions during continental flood-basalt genesis: Geochemical evidence from cretaceous magmatism in southern Brazil. Contributions to Mineralogy and Petrology. 1999;**137**:147-169. DOI: 10.1007/s004100050588

[39] Gibson SA et al. Late cretaceous rift-related upwelling and melting of the Trindade starting mantle plume head beneath Western Brazil. Contributions to Mineralogy and Petrology. 1997;**126**:303-314. DOI: 10.1007/s004100050252

[40] Thompson RN et al. Migrating cretaceous–eocene magmatismin the Serra do mar alkaline province, SE Brazil: Melts from the deflected Trindade mantle plume? Journal of Petrology. 1998;**39**(8):1493-1526. DOI: 10.1093/petroj/39.8.1493

[41] Ulbrich H, Gomes CDB. Alkaline rocks from continental Brazil. Earth-Science Reviews. 1981;**17**(1-2):135-154. DOI: 10.1016/0012-8252(81)90009-X

[42] Oliveira AL et al. Green core clinopyroxenes from Martin Vaz archipelago Plio-Pleistocenic alkaline rocks, South Atlantic Ocean, Brazil: A magma mixing and polybaric crystallization record. Journal of South American Earth Sciences. 2021;**105**:102951. DOI: 10.1016/j.jsames.2020.102951

[43] Santos AC et al. Martin Vaz archipelago: The youngest magmatism

in the Brazilian territory. In: Meso-Cenozoic Brazilian Offshore Magmatism. London: Academic Press, Elsevier; 2022. pp. 391-432. DOI: 10.1016/B978-0-12-823988-9.00004-6

[44] Santos AC et al. Vitória-Trindade seamounts: Undersaturated alkaline series evolution from an enriched metasomatized source. In: Meso-Cenozoic Brazilian Offshore Magmatism. London: Academic Press, Elsevier; 2022. pp. 293-336. DOI: 10.1016/B978-0-12-823988-9.00004-6

[45] Maia TM et al. First petrologic data for Vitória seamount, Vitória-Trindade Ridge, South Atlantic: A contribution to the Trindade mantle plume evolution. Journal of South American Earth Sciences. 2021;**109**:103304. DOI: 10.1016/j.jsames.2021.103304

[46] Maia TM et al. Abrolhos magmatic province petrogenesis and its link with the Vitória-Trindade ridge, Southeast Brazilian margin, South Atlantic Ocean. Journal of South American Earth Sciences. 2022;**120**:104075. DOI: 10.1016/j.jsames.2022.104075

[47] Monteiro LGP et al. Trindade Island: Evolution of the geological knowledge. In: Meso-Cenozoic Brazilian Offshore Magmatism. London: Academic Press, Elsevier; 2022. pp. 337-389. DOI: 10.1016/j.jsames.2022.104075

[48] Rego CAQ et al. Davis Bank geodynamic context, South Atlantic Ocean: Insights into the Vitória-Trindade ridge evolution. Journal of South American Earth Sciences. 2021;**112**:103620. DOI: 10.1016/j.jsames.2021.103620

[49] O'Connor JM, Duncan RA. Evolution of the Walvis ridge-Rio Grande rise hot spot system: Implications for African and South American plate motions over plumes. Journal of Geophysical Research: Solid Earth. 1990;**95**(B11):17475-17502. DOI: 10.1029/JB095iB11p17475

[50] Bongiolo EM et al. Geochemical modeling and Nd–Sr data links nephelinite–phonolite successions and xenoliths of Trindade Island (South Atlantic Ocean, Brazil). Journal of Volcanology and Geothermal Research. 2015;**306**:58-73. DOI: 10.1016/j.jvolgeores.2015.10.002

[51] de Jesus JVM et al. Petrogênese do Banco Davis, Cadeia Vitória-Trindade, Atlântico Sul: o Papel de Voláteis (H_2O e CO_2) na Evolução Magmática do Banco Davis. Anuário do Instituto de Geociências. 2019;**42**(3):237-253. DOI: 10.11137/2019_3_237_253

[52] Santos AC. Petrography, Lithogeochemistry and 40 Ar/39 Ar Dating of the Seamounts and Martin Vaz Islands-Vitória-Trindade Ridge [Thesis]. Rio de Janeiro: Universidade do Estado do Rio de Janeiro; 2013. 115 p. CDU 552(815.2)

[53] Santos AC. Petrology of Martin Vaz Island and Vitoria Trindade Ridge Seamounts: Montague, Jaseur, Davis, Dogaressa and Columbia. Trace Elements, 40Ar/39Ar Dating and Sr and Nd Isotope Analysis Related to the Trindade Plume Evidences. Rio de Janeiro: Faculdade de Geologia, Universidade do Estado do Rio de Janeiro; 2016. Available from: http://www.bdtd.uerj.br/handle/1/7023

[54] Santos AC et al. Pleistocene alkaline rocks of Martin Vaz volcano, South Atlantic: Low-degree partial melts of a CO_2-metasomatized mantle plume. International Geology Review. 2019;**61**(3):296-313. DOI: 10.1080/00206814.2018.1425921

[55] Santos AC et al. Compiled potential field data and seismic surveys across the

eastern Brazilian continental margin integrated with new magnetometric profiles and stratigraphic configuration for Trindade Island, South Atlantic, Brazil. International Geology Review. 2019;**61**(14):1728-1744. DOI: 10.1080/00206814.2018.1542634

[56] Siebel W et al. Trindade and Martín Vaz Islands, South Atlantic: Isotopic (Sr, Nd, Pb) and trace element constraints on plume related magmatism. Journal of South American Earth Sciences. 2000;**13**(1-2):79-103. DOI: 10.1016/S0895-9811(00)00015-8

[57] Menzies M, Murthy VR. Mantle metasomatism as a precursor to the genesis of alkaline magmas-isotopic evidence. American Journal of Science. 1980;**280**:622-638. DOI: 10.1130/SPE215-p47

[58] Pilet S et al. What petit-spot volcanoes tell us about the lithosphere-asthenosphere boundary? In: American Geophysical Union, Fall Meeting 2015, abstract id. T34C-04. San Francisco: AGUFM; 2015. T34C. 4 P

[59] Pilet S et al. Monte Carlo simulations of metasomatic enrichment in the lithosphere and implications for the source of alkaline basalts. Journal of Petrology. 2011;**52**(7-8):1415-1442. DOI: 10.1093/petrology/egr007

[60] Pilet S, Baker MB, Stolper EM. Metasomatized lithosphere and the origin of alkaline lavas. Science. 2008;**320**(5878):916-919. DOI: 10.1126/science.115656

[61] Pilet S et al. Short-term metasomatic control of Nb/Th ratios in the mantle sources of intraplate basalts. Geology. 2004;**32**(2):113-116. DOI: 10.1130/G19953.1

[62] Pilet S et al. The metasomatic alternative for ocean island basalt chemical heterogeneity. Earth and Planetary Science Letters. 2005;**236**(1-2):148-166. DOI: 10.1016/j.epsl.2005.05.004

[63] Niu Y et al. A trace element perspective on the source of ocean island basalts (OIB) and fate of subducted ocean crust (SOC) and mantle lithosphere (SML). Episodes Journal of International Geoscience. 2012;**35**(2):310-327. DOI: 10.18814/epiiugs/2012/v35i2/002

[64] Fodor RV, Hanan BB. Geochemical evidence for the Trindade hotspot trace: Columbia seamount ankaramite. Lithos. 2000;**51**(4):293-230. DOI: 10.1016/S0024-4937(00)00002-5

[65] Almeida FFM. Ilhas oceânicas brasileiras e suas relações com a tectônica atlântica. Terræ Didatica. 2006;**2**(1):3-18. Available from: http://www.ige.unicamp.br/terraedidatica/

[66] Santos ACD et al. Geology of Martin Vaz Island, South Atlantic, Brazil. Journal of Maps. 2015;**11**(2):314-322. DOI: 10.1080/17445647.2014.936913

[67] Cordani UG. Idade do vulcanismo no Oceano Atlântico Sul. Boletim IGA. 1970;**1**:09-75. DOI: 10.11606/issn.2316-9001.v1i0p09-75

[68] Pires GLC et al. New 40Ar/39Ar ages and revised 40K/40Ar* data from nephelinitic–phonolitic volcanic successions of the Trindade Island (South Atlantic Ocean). Journal of Volcanology and Geothermal Research. 2016;**327**:531-538. DOI: 10.1016/j.jvolgeores.2016.09.020

[69] Santos AC et al. Martin Vaz island geochronology: Constraint on the Trindade mantle plume track from the youngest and easternmost volcanic episodes. Journal of South American Earth Sciences. 2021;**106**:103090. DOI: 10.1016/j.jsames.2020.103090

[70] Ferrari AL, Riccomini C. Campo de esforços plio-pleistocênico na Ilha de Trindade (Oceano Atlântico Sul, Brasil) e sua relação com a tectônica regional. Revista Brasileira de Geociencias. 1999;**29**(2):195-202. DOI: 10.25249/0375-7536.199929195202

[71] Jesus JVM. Petrologia das rochas do Banco Davis, Cadeia Vitória-Trindade, Atlântico Sul.: dados de química de rocha total, DRX e Microssonda. Rio de Janeiro, 109 p: Monografia–Faculdade de Geologia, Universidade do Estado do Rio de Janeiro; 2018

[72] Quaresma GOA. Geoquímica Isotópica Sr-Nd-Pb e Geocronologia 40Ar/39Ar do Banco Submarino Davis. In: Cadeia Vitória-Trindade: Possíveis Implicações com a Movimentação da Placa Sul-Americana durante o Mioceno. Rio de Janeiro: Capstone Project, Universidade Estadual do Rio de Janeiro; 2019

[73] Alberoni AAL et al. The new digital terrain model (DTM) of the Brazilian continental margin: Detailed morphology and revised undersea feature names. Geo-Marine Letters. 2020;**40**(6):949-964. DOI: 10.1007/s00367-019-00606-x

[74] Palma JJC. Fisiografia da área oceânica. In: Geologia do Brasil. Vol. 501p. Brasília: Ministério das Minas e Energia, Departamento Nacional de Produção Mineral; 1984

[75] Motoki A, Motoki KF, Melo DP. Submarine morphology characterization of the Vitória-Trindade chain and the adjacent areas, state of Espírito Santo, Brazil, based on the predicted bathymetry of the TOPO version 14.1. Revista Brasileira de Geomorfologia. 2012;**13**(2):151-170. DOI: 10.20502/rbg.v13i2.195

[76] Krawez M et al. Building dense reflectance maps of indoor environments using an rgb-d camera. In: 2018 IEEE/RSJ International Conference on Intelligent Robots and Systems (IROS), Madrid, Spain. 2018. pp. 3210-3217. DOI: 10.1109/IROS.2018.8594107

[77] Ruisi Z, Min Z, Jianping C. Study on geological structural interpretation based on worldview-2 remote sensing image and its implementation. Procedia Environmental Sciences. 2011;**10**:653-659. https://doi.org/10.1016/j.proenv.2011.09.105

[78] Hunt G. Spectral signatures of particulate minerals in the visible and near infrared. Geophysics. 1977;**42**:501-513. DOI: 10.1190/1.1440721

[79] Geoffroy L, Bergerat F, Angelier J. Tectonic evolution of the Greenland-Scotland ridge during the Paleogene: New constraints. Geology. 1994;**22**(7):653-656. DOI: 10.1130/0091-7613(1994)022<0653:TEOTGS>2.3.CO;2

[80] Jones OT. The dynamics of faulting and dyke formation: With applications to Britain. Nature. 1942;**149**:651-652. DOI: 10.1038/149651b0

[81] Ramsay J, G, Huber MI, Lisle RJ. Folds and fractures. In: The Techniques of Modern Structural Geology. Vol. 2. London: Academic Press, Elsevier Science; 1986. pp. 309-700. ISBN-10.0125769229

[82] Tchalenko JS. Similarities between shear zones of different magnitudes. Geological Society of America Bulletin. 1970;**81**(6):1625-1640. DOI: 10.1130/0016-7606(1970)81[1625:SBSZOD]2.0.CO;2

[83] Wilcox RE, Harding TPT, Seely DR. Basic wrench tectonics. AAPG Bulletin. 1973;**57**(1):74-96. DOI: 10.1306/819A424A-16C5-11D7-8645000102C1865D

[84] Bingham CJTAOS. An antipodally symmetric distribution on the sphere.

The Annals of Statistics. 1974;**2**:1201-1225. DOI: 10.1214/aos/1176342874

[85] Barão LM et al. Deformational structures developed in volcanic sequences as a product of tectonic adjustments in the South Atlantic Ocean. Journal of South American Earth Sciences. 2020;**104**:102812. DOI: 10.1016/j.jsames.2020.102812

[86] Mendiguren JA, Richter FM. On the origin of compressional intraplate stresses in South America. Physics of the Earth and Planetary Interiors. 1978;**16**(4):318-326. DOI: 10.1016/0031-9201(78)90070-5

[87] Artyushkov EV. Stresses in the lithosphere caused by crustal thickness inhomogeneities. Journal of Geophysical Research. 1973;**78**(32):7675-7708. DOI: 10.1029/JB078i032p07675

[88] Korenaga J. Initiation and evolution of plate tectonics on earth: Theories and observations. Annual Review of Earth and Planetary Sciences. 2013;**41**:117-151. DOI: 10.1146/annurev-earth-050212-124208

Chapter 2

Formation and Evolution of the Sarmatia Earth's Crust (East European Craton): Evidence from the Dnipro-Donets Paleorift

Oleksii Bartashchuk

Abstract

The East European Craton tectonics was studied based on the reconstruction of the stress field of Sarmatia. The Sarmatia internal plate field is influenced by induction stress from the movements of the Arabia and Scythia plates and the North Atlantic Ridge spreading. Stress from the plate boundaries is transferred inward and absorbed by the movable belts. The role of movable belts in the evolution of the lithosphere is considered on the basis of the Dnieper-Donet Paleorift. At the Hercynian-Alpine collision stage, the rift basin experienced inversion uplift and folding. The change in deformation modes was initiated by displacement of the geodynamic axes by 15° counterclockwise during the epoch with a total of 60° in Phanerozoic. The rift structure was deformed by longitudinal bending on diagonal strike-slips of opposite kinematics and longitudinal elongation on one-sided strike-slips or along the diagonal strike-slips. The framework of the deformations is made up of diagonal sutures form the "Longitudinal Extrusion Orocline" and the "Transverse Spreading Fan". Consequently the West Donets Covered Folded Region and the Donets Foldbelt were formed and the Paleorift has turned into a movable foldbelt.

Keywords: Earth's crust, stress, deformation, inversion, East-European Craton, Sarmatia, Dnieper-Donets Paleorift, West Donets Covered-Folded Region

1. Introduction

The tectonic evolution of the lithosphere and Earth's crust is characterized by its stages and cyclic nature and occurs permanently in a variable stress field, favoring one of the elementary deformation modes of compression, tension, or horizontal shifting [1–4]. The key factor in the structural-kinematic evolution of the continental crust is the reactivation of tectonic faults along regular directions of the Planetary fracture grid [2, 3]. During crust evolution, individual tectonic frameworks are formed, controlling various age deformation planes of the original structure. The movements of activated rock masses along deformation frameworks initiate the transformation from block-faulted tectonics to fold-dislocated due to internal redistribution of masses and deformation structure formation [4]. Faults are re-mobilized in accordance with the nature of the actual stress field. In another geodynamic environment, a new deformation mode is activated, according to which the structural-dynamic manifestation of faults in the crust, the direction, and kinematics of massif and block movements change [5]. The Planetary grid is the basis of the global deformation framework, whose dynamics and kinematics are renewed through fault re-mobilization in a variable stress field. The invariance of deformation regimes along established directions of the grid is conditioned by spatial-temporal inversions of stress field parameters [2, 3, 6].

Tectonic deformations disrupt the isotropy of the lithosphere by dividing it into tectonic blocks and plates through the formation of fractures, faults, linear zones of fracture-dislocations, and mobile and collision belts. The character of the divisibility is conditioned by the anisotropy of the lithosphere, temporal changes of Planetary geodynamic processes, and directions of the Planetary fracture network. Faults disrupt the consolidated crust through fracturing and influence the character of crust structural evolution due to mobility and the ability to remobilize [2–6]. Faults consist of several different age shear deformation zones of approximate azimuth orientation, so their structure is determined by the genetic types of fractures that compose them [3, 5]. Faults form deformation frames in the directions of the Planetary grid [2] during their evolution. The grid has a steady orientation in geochronology relative to the figure of the Earth [2, 3]. The horizontal axes of tension ($\sigma 1$) and extension ($\sigma 3$) are located according to modern meridians and parallels, but in earlier geological epochs were rotated at angles multiple to 15 [2, 3, 5]. The periodicity of stress field inversions is due to the change in Planetary rotation parameters, so the periodicity of changes in deformation regimes is close to the geological epoch [2, 3]. According to the new mode, the genetic type of faults, directions, and kinematics of movements in different age structural surfaces change [2, 6].

The internal plate field of the East-European Craton (EEC) is most influenced by induction stress from the movements of the Arabia and Scythia continental plates and the North Atlantic Ridge spreading [7–10]. To understand the deformation mechanism in areas of lithosphere plate collision, the direction and kinematics of displacements of rock masses, located between the plates, are usually reconstructed. Two vertical stress outflow directions: pulling down the lithosphere edge by a descending convection current during subduction, or squeezing upwards with uplift and fold formation, have a vivid structural manifestation in the modern relief and are relatively easily determined. However, within continental plates, significant horizontal movements occur within collision belts that are decisive for structure formation. The structural results of lateral displacements in intra-plate collision belts are not always reflected at the surface and are less confidently distinguished based on the analysis of structural patterns of deformations [2–7]. The result of deformations by the mechanism of lateral squeezing of masses along the belts is the strike-slip structures as a diagnostic indicator [4, 5, 7]. The division of plates into relatively independent micro-plates by movable belts contributes to geodynamic independence and increases the overall plate mobility. Intra-continental rift belts are movable zones capable of absorbing and accumulating stress. When studying the mechanisms of intra-plate deformations, the general kinematics of movements of distant collision belts are taken into account [7–10].

The core of the EEC is composed of three segments: Fenno-Scandia, Sarmatia, and Volga-Uralia (**Figure 1**) [11]. The internal tension field and deformations of the Sarmatia crust are most influenced by induction stress from the movements of the Arabian and Scythian continental plates and the North Atlantic Ridge spreading [7–10].

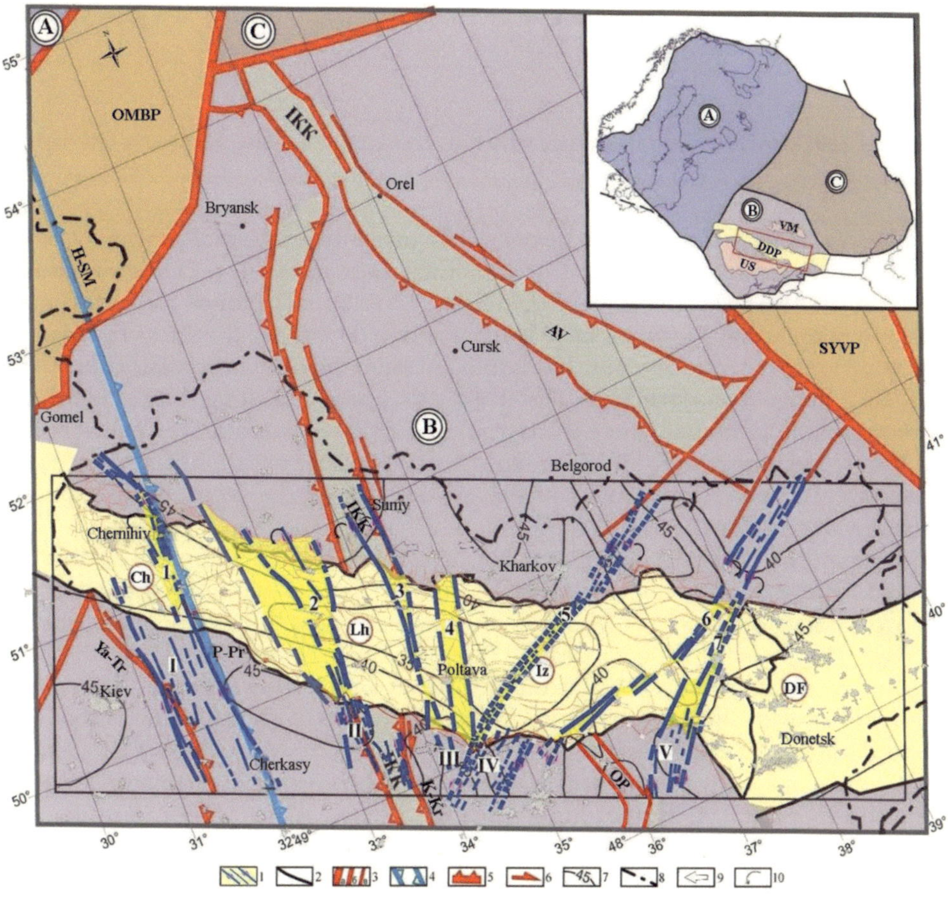

Figure 1.
Position of the Dnieper-Donets Paleorift on the Sarmatia plate and the Eastern European Craton (inset), by [11, 12] with additions, by [13–15]. Symbols: 1—intersegment zones; 2—Paleorift boundaries; 3—faults: a—interplates, b—intermega-blocks, c—interblocks; 4—tectonic seam (H-SM-Kherson-Smolensk) and neo-tectonics strike-slip zones; 5—volcano belts (OMBP—Osnytsko-Mikashevychsky; SYVP—Stavropol-Yertil); 6—kinematic of displacement; 7—depth of the Moho surface (km); 8—state border; 9–10—directions of general and rotational movements of rocks. Letters on the maps: ABC—plates of the Eastern European Craton: A—Fenno-Scandia, B—Sarmatia, C—Volga-Uralia; tectonic structure: VM—Voronezh Massif, DDP—Dnieper-Donets Paleorift, US—Ukrainian shield. Segments of the Dnieper-Donets Paleorift: Ch—Chernihiv, Lh—Lokhvytsa, Iz—Izium; DF—Donets Foldbelt. Inter-block zones of the Ukrainian shield: I—Kropyvnitskiy; II—Ingulets-Kryvyi Rih-Kremenchug; III—Kamiaynsk; IV—Odessa; V—Central Priazov-Sloviyanogirsk. Suture belts and zones: IKK—Ingulets-Kryvyi Rih-Krupets`, AV—Alekseevka-Voronezh belts; Ya-Tr—Yadliv-Traktemyriv, K-Kr—Kryvyi Rih-Kremenchug, OP—Orikhiv-Pavlograd, by [2, 3, 12]. Frame of inversion deformations: 1—Kropyvnitskiy; 2—Ingulets-Kryvyi Rih; 3—Kamiaynsk; 4—Verkhovtsy-L'hov; 5—Kolomak-Kobeliaky; 6—Balakliya-Synelnykove; 7—Priazov-Slavyanogirsk, by [2, 13–15].

During the pre-rift stage of evolution, due to lithosphere anisotropy, the Sarmatia was divided into West and East micro-plates and smaller blocks by ancient collision suture and seams [12]. At the rift stage, through the formation of the Dnieper-Donets Paleorift (DDP), a new division of the Sarmatia lithosphere occurred, isolating the massifs of the Ukrainian Shield (USh) and Voronezh Anticlise (VA). During the Hercynian, Cimmerian, and Alpine platform activation, the rift basin underwent uplift, folded, and over-thrust deformations and transformed into a collision foldbelt [6, 8–10, 16]. The problems of the formation and evolution of the Earth's crust of the

EEC are discussed below on the basis of the latest data on tectonics and geodynamics of the evidence from Sarmatia and DDP.

2. Stress and deformation sources of Earth's crust of the Eastern European Craton

The stress field of the Earth is the result of a constant horizontal alignment of the maximum stress axis σ1 on continental lithosphere plates [17]. The established correlation between the direction of the stress axis and the vector of plate movement speed is evidence that the forces of plate interaction determine the stress field in intra-plate areas [17, 18]. The EEC has homogeneous collision compression due to its relatively homogeneous structure. The stress axis is oriented along the meridian, with the exception of Sarmatia, where it has a north-northwest orientation, associated with lithosphere anisotropy [12]. The stress field structure within EEC is determined by the sum of forces from various sources and depends on the degree of lithosphere anisotropy of constituent plates.

Contemporary and ancient stress fields are characterized by different levels of crust deformations. Contemporary field tensors are determined instrumentally [3]. Both strong and weak stresses, which do not cause residual deformation of rocks, are analyzed. The tensor of the ancient field is restored using the mechanical indicators of rocks and minerals, so only those stresses are determined whose magnitude exceeds the limit of irreversible deformations. The consequences of ancient deformations are observed only in the near-surface parts of the crust, but modern stresses can cover the entire upper consolidated crust with a thickness of 20–25 km. The deviator component of the field is taken into account—the excess stress, which is formed under the influence of certain forces in addition to litho-static (gravy) loading. If the stress field is stable for the upper part of the crust (20–25 km), the stress axes orientation on the horizontal with depth does not change [18]. The stress magnitude increases in consolidated rocks, so the stress deviator component is most often manifested in the basement and usually absent in the sedimentary cover of the platforms. The exception is the sheared folded mantles of intra-plate basins, where the stress field can be independent of the autochthonous platform they overlie [19].

Sources of stress causing deformation of the lithosphere are considered [15]: Planetary rotational forces, caused by changes in the parameters of the planet's rotation and the Earth-Moon system; load forces on the Earth's surface of the sedimentary thickness, tectonic and volcanic covers, negative and positive structures; local gravitational inhomogeneities, caused by variations in the density of mountain rock masses; forces arising during uneven bending of layers during heating/cooling of the rheological layered lithosphere; forces associated with the movements of lithosphere plates, including membrane forces, initiated by the movement of plates on the Geoid with variable curvature ("Plate driving forces").

Strong stresses (up to 90 MPa) inside the plates can create mountain ridge loads [20]. However, the static pressure of excess masses creates anomalous fields only within the height of relief and does not affect the regular field of lithospheric plates. Stress inside the plates is influenced by movements at subduction borders: slab pull forces and trench suction [21]. However, the overall action of the plates is limited to the area adjacent to the subduction zone, so it does not significantly affect the upper part of the crust inside the lithospheric plates. The effect of drag and resistance forces that can act at the base of lithosphere plates ("Drag forces mantle drag") is also insignificant [22]. A significant source of stress is the pressure of spreading oceanic ridges. Evidence of this is the intra-plate stress in the east of North America and in the northwest of Europe as a result of crustal divergence in the Mid-Atlantic Ridge [17, 18]. Regarding the influence of continental rifting, it has been established that against the background of crustal extension stages, episodes of compression and inversion uplift occur due to stress inversion in the lower layers of the lithosphere. Tectonic and sedimentary inversion of the rift is accompanied by changes in the thickness of the basin's sedimentary cover without changing the overall movement of lithosphere plates [23]. Therefore, to model the evolution of the rift basin, the factors of dynamic interaction between the forces of the distant stress field from movements in collision belts and the loading/unloading of the basin structure through denudation and under the influence of static pressure of relief heights are taken into account. During the pauses between stress episodes up to 30 MPa, rift basins experience an uncompensated inversion of depths (up to 2 km) due to the cessation of debris inflow from the rising banks. This creates an additional force source for the formation of tectonic rifts during activation phases against the general divergence of continents.

Thus, Planetary and local forces act together, but the decisive contribution to the stressed state of the lithosphere is made by Planetary forces, the rest of the forces provide secondary variations. Collision stress propagates inside the plate for hundreds of km in EEC. The internal plate field is most influenced by induction stress from the movements of the Arabian and Scythian continental plates and the North Atlantic Ridge spreading. The Alpine-Himalayan collision belt, as a global structure-concentrator of stresses and deformations, has a greater influence on the state of the modern stress field in Earth's crust of the EEC [7–10].

3. Formation and evolution of the Sarmatia Earth's crust

3.1 Anisotropy of lithosphere of the Sarmatia

The DDP is an element of the Sarmatia-Turan intraplate movable belt of the Eurasian Platform, extending from the Pripyat depression trough to the mountain-folded Tian-Shan [7]. The DDP crosses the Sarmatia obliquely, dividing the Archean-Proterozoic crystalline massifs of the USh and the VA (**Figure 1**) [2, 11, 12]. The structure of the Paleorift consists of an early Riphean Trog in the basement, a Devonian Graben and the overlying upper Paleozoic over-rift depression and Mesozoic-Cenozoic platform syncline (**Figure 2**) [13, 24]. The boundary faults of the Graben reach the Moho, form the tectonic shoulders, and control the early Riphean and Late Devonian rift complexes [24]. The Graben axis is convexly curved to the south and changes its stretch from sub-latitudinal to latitudinal in the southeast direction. In the same direction, the depths of the surface of the Archean-Proterozoic folded foundation increase from 5 to 20–22 km (**Figures 1** and **3**) [12, 13]. The Graben structure is diagonally superimposed on the pre-rift meridian seams (Kherson–Smolensk and Ingulets-Krivoy Rog, Donetsk–Bryansk) and crust-mantle suture (Odessa-Talniv, Ingulets-Bryansk, Verkhovtsy-L'gov, Odessa, and Central-Azov-Slovyansk) (**Figure 1**). The sub-meridian tectonic seams and sutures are fan-shaped and divided into component crust-mantle faults from the USh in the northern direction on the slopes of the VA (**Figures 1** and **3**) [25]. The transverse frame divides the structure from west to east into the Chernihiv, Lohvitsa, Izium, West Donets (segments), and the Donets Foldbelt (DF) known as the Paleo-basin "Donbas" (**Figure 3**). The Ingulets-Krivoy Rog-Krupets tectonic seam divides the VA into the Bryansk (West) and Kursk (East) micro-plates, and the USh into the West and East micro-plates. The Verkhovtsy-Lgov fault is the central lineament of the Ingulets-Krivoy Rog-Krupets seam, which breaks the Paleorift almost in half (**Figures 1** and **3**) [25]. Along the fault, the axis of tectonic symmetry formed, which separates the West and East micro-plates of Sarmatia of different structure, thickness and substance composition of the lithosphere layers, and opposite kinematics of displacements of rock [13]. The lithosphere under the Graben thins southeastward from 35 to 42 km in the Chernihiv to 15–25 km in the Lohvitsa and Izium segments, and thickens to 25–30 km at the border with the DF (**Figure 4**) [12]. The density of the crust at the Moho boundary increases southeastward from 3.12 g/cm^3 in.

Chernihiv to 3.18 g/cm^3 in the Lohvitsa and from 3.18 to 3.20 g/cm^3 in the Izium to 3.36 g/cm^3 in the DF (**Figure 5**). The most powerful lithosphere up to 280 km lies under the Chernihiv, but under the Lohvitsa is a dome-like rift relic at the base of the lithosphere at a depth of 120 km. The base of the lithosphere rises from the Izium to DF to 80 km, which is due to the opening of the Devonian rift basin to the southeast in the Tethys Ocean [8–10].

Due to structure-material differentiation in the mantle at depths of 50–250 km, zones of anomalous longitudinal seismic wave speeds were formed (**Figure 4**) [12]. In the zone of reduced speed under the Lohvitsa and the east of the Chernihiv, a "layering" of speed anomalies is localized. In the lithosphere under it, at depths of 70–130 km, a layer of anomalously high speed was formed through the intrusion of

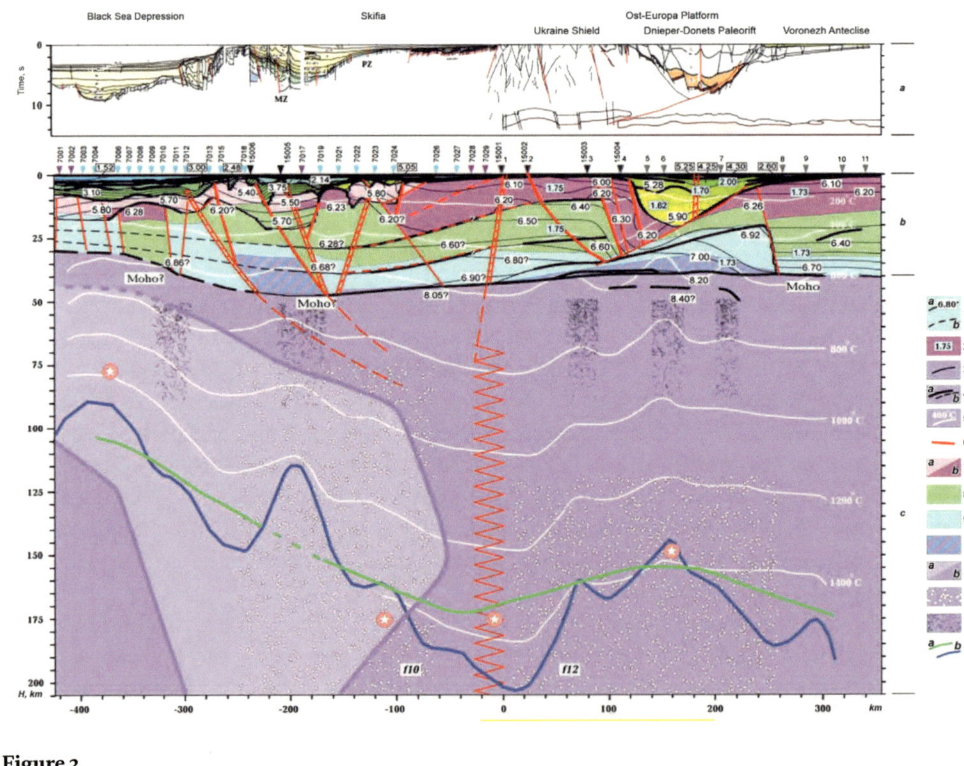

Figure 2.
Sections of the lithosphere East Europe Craton along the DOBRE fraction'99/DOBRE-2 profile, by [24]. Symbols: a—sedimentary cover, b—consolidated crust, c—mantle; 1—P-wave speed, 2—Vp/Vs, 3—reflection horizons, 4—Moho, 5—isotherm, grad C, 6—faults, 7, 8, 9—Earth's crust: 7—upper (low and regular density), 8—middle, 9—bottom; 10—crust-mantle mix, 11—low and speed mantle, 12—deep fluids, 13—thermal anomalies, 15—roof of the asthenosphere: geothermal and seismic.

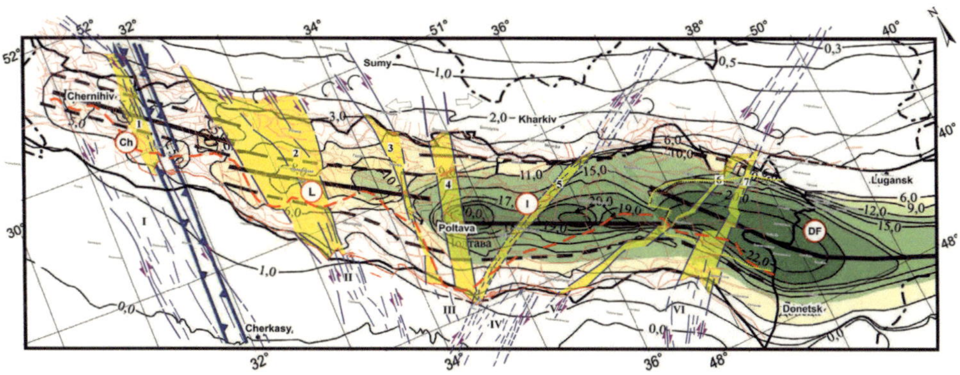

Figure 3.
Scheme of the Precambrian crystal basement relief of the Dnieper-Donets Paleorift, by [13]. Symbols in ***Figure 1****.*

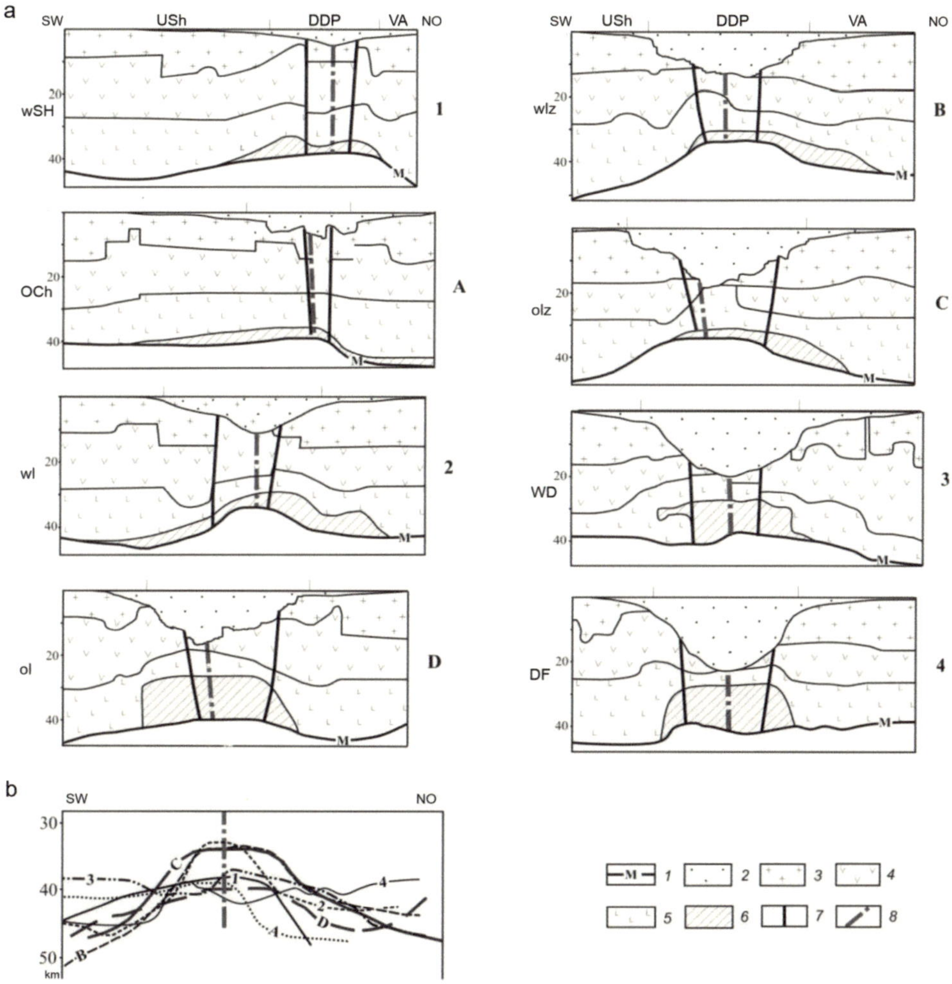

Figure 4.
The deep structure of the Sarmatia Earth's crust on sections across the Dnieper-Donets Paleorift, by [13]. Symbols: 1—the Moho boundary; 2–6—layers of the Earth's crust: 2—sedimentary, 3—granite, 4—diorite, 5—basalt, 6—crust-mantle mixture; 7—edge faults of the Paleorift; 8—axis of the Paleorift.

mantle matter from under the massif of the USh [12, 24]. The mantle subducted slab was facilitated by a shallow tilt of the northern side fault plane, which type changes with depth from a down-throw to a subduction-throw (**Figure 2**) [25]. In the west of the Chernihiv, the consolidated crust is made up of a "diorite layer," and the subcrustal mantle has increased longitudinal seismic wave speeds (**Figure 4**). In the east, a typical "granite-free" rift crust is developed, and magnetic bodies are localized in the densified "diorite" and "basalt layers." The relief of the Moho boundary coincides with the relief of the lithosphere base. In the Lohvitsa, the crust is "granite-free," and the "basalt layer" also decreases in thickness due to the thickening of the "crust-mantle mixture" layer. In the roofs of the "basalt layer," "crust-mantle mixture," the bases of crust and seismic lithosphere, a dome-like rise is formed, which is a relic of the early Riphean rift.

The western part of the Izium has a "granite-free" crust with a thinned "diorite layer," the "crust-mantle mixture" is ubiquitous (**Figure 4**). "Diorite" and "basalt" layers are densified and magnetic. Here a gradient rise of the Moho boundary and a shallow rise of the lithosphere base were formed. In the subcrustal mantle, a lens-shaped body with a density of 3.50 g/cm^3 is localized (**Figure 5**). In the east part of the Izium, the crust contains a thin "diorite layer," and a rise is traced in the roof of the "basalt layer." The "crust-mantle mixture" forms a lens of abnormally high density (3.20 g/cm^3) with a thickness of 12 km at the Moho boundary. Under the Moho boundary at depths of 60–80 km, the upper mantle is densified to 3.50 g/cm^3. In the West Donets, the thinning of the "diorite layer" under the axial part of the Graben is associated with the maximum rise of the roof of the "basalt layer." On the west slopes of the DF, sedimentary cover in the basement of the Graben lies on the "basalt layer." The "crust-mantle mixture" layer reaches the maximum thickness of 16 km. On the Moho boundary in the Central-Azov-Slovyansk suture, a flexure of 5 km amplitude was formed (**Figures 3–5**).

Anisotropy of the Sarmatia lithosphere under the DDP is manifested from the surface to the base in variations in the composition and thickness of Earth's crust layers, relief of the Moho, the lithosphere basement, and the consolidated crust (**Figures 2–5**). The longitudinal structural-material differentiation of the lithosphere is in the structure differences of the West and East micro-plates. An inhomogeneity in the orientation of the shear wave vector was established [12], which is considered

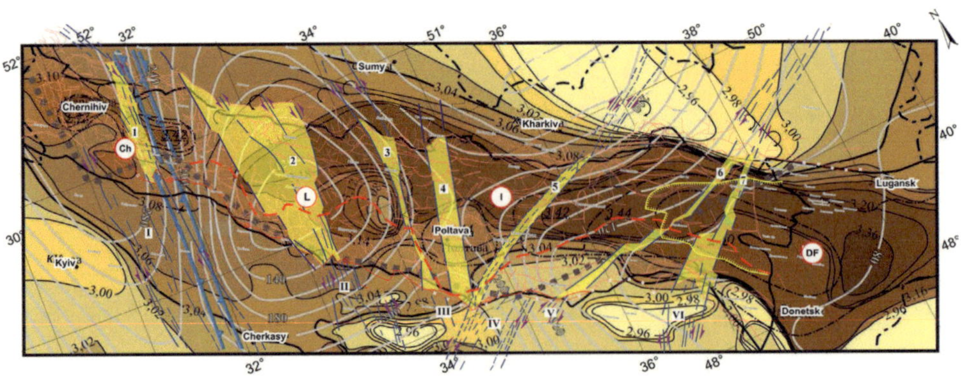

Figure 5.
*Scheme of the sole of the Dnieper-Donets Paleorift lithosphere and the density on the Moho boundary, by [13]. Symbols in **Figure 1**. Notations: The dashed line is the north boundary of the high-velocity horizontally in the Golitsyn-Heiko layer.*

evidence of structural differences, conditioned by the "built-in" ("frozen") anisotropy of the lithosphere [25].

3.2 Geodynamic settings of the Sarmatia Earth's crust deformations

The division of the EEC by intra-plate collision and movable belts into three lithosphere plates (Fenno-Scandia, Sarmatia, and Volga-Uralia) provides them with relative geodynamic independence and increases the overall and intra-platform tectonic mobility. At around 2.0 Ga subduction the oceanic plate, which separated Sarmatia from Volga-Uralia, has led to the collision with Fenno-scandia. The accretion of the two continental plates was completed at around 1.75 Ga. The Sarmatia is a stable Archean Craton, created by Late Archean-early Paleoproterozoic fusion of several different ages micro-plates ranging from 3.8 to 2.8 Ga [24]. Neoarchean-Proterozoic crust-mantle sutures intersect the 180–260 km thick Sarmatia lithosphere and sink into the mantle to 100–200 km (**Figures 1–3**). Sutures form a tectonic framework that controls the processes of Earth's crust structural evolution due to repeated reactivation in a variable stress field. Within the USh, the DDP and the VA, displacements with horizontal amplitudes of tens of km occurred along this framework. In the Precambrian, the west and east parts of both crystalline massifs developed as separate micro-plates and were located far apart from each other. During the Vendian and Phanerozoic, the North pole moved from equatorial latitudes to the present position, and both micro-plates became part of Sarmatia [2, 3, 8–10] due to a 90° clockwise rotation of the EEC [3].

The model of the Precambrian evolution of Sarmatia covers several stages of collision with the subduction of continental crust and spreading with the divergence of the West and East micro-plates [3]. At Paleoproterozoic, collision of the Middle Dnieper and Azov Archean blocks were directed in the northeast and southwest of the current coordinates. Three continental (Rosin-Bug, Middle Dnieper, Azov) and two sub-oceanic micro-plates (Kryvyi Rih-Odessa, Dnieper-Azov) participated in the convergent process. Subduction of oceanic micro-plates under the Middle Dnieper and Azov micro-continents occurred with the closure of the Kryvyi Rih-Odessa paleo-basin. The movement of plates over the mantle plume ended with the form of the Holovaniv-Inhulets-Kryvyi Rih collisional suture. Due to the intrusion of the second plume, the Western and Eastern micro-plates diverged from the collision suture, forming a Trog between the Pervomaisk and West Inhulets collision zones. After the plume was extinguished, the second convergence of the West and East micro-plates of the Talne and Kryvyi Rih-Kremenchuk fault zones was formed. Under the influence of the third plume, the Holovaniv-Inhulets-Kryvyi Rih collision suture diverged, forming the Holovaniv and Inhulets-Kryvyi Rih sutures. The processes of crust convergence and divergence conditioned the anisotropy of the Sarmatia lithosphere through the formation of movable suture zones.

The model of tectonic and magmatic evolution of the DF includes several stages inherent to continental rifting [14]. The Riphean stage is associated with crustal stretching, the formation of a chain of Trogs, and the Sambets basaltic magma complex with gold-silver and copper mineralization. After the Early Paleozoic stabilization, rifting began in the Devonian with the reactivation of the old rift structure and the formation of a Graben belt. Volcano-plutonic magma rocks are composed of the Azov titan-pyroxenite and differentiated volcano complex from diamond-bearing alkali picrite to rhyolite. The synclinal stage in the Carboniferous involved submersion and the deposition of a coal-carbonate formation. Against the background of platform activation, the Anastasievka magmatic complex was formed, composed of gabbroid dikes and a volcano from basalt to alkaline rhyolite. After its completion, the Permian-Early Triassic stage of inversion uplift of the basin took place. It was accompanied by intrusions of granites, diorites of the shonkinite-monsonite-plagiophyr phase of the South Donbas complex. In the late Triassic, the stage of activation began, caused by movements and pressure from the south of the Caucasian micro-plates. It is associated with intrusions of andesites of the Nesvitaevka and lamprophyre Mius-Kerchik complexes. The stage of stabilization set in with the transition to platform mode at the boundary of the Cretaceous and Paleogene before the neo-tectonic activation. Fading tectonic movements were accompanied by diorite-dacitic dikes, low-temperature hydro-thermalites, and crushing of Lower Cretaceous rocks.

The kinematic model of rifting presupposes the formation according to the "athermal" mechanism through the elastic fracture of the "cold" continental Earth's crust [14]. At the pre-rift collision stage in the Late Proterozoic, in the north-western diagonal faults at azimuths 291–312°, and 315–339°, a linear "germinal" strike-slip zone was formed. At the early Riphean in the shear stress field, movements along the zone initiated the formation of a chain of Trogs by the Pull-apart basin mechanism. At the Devonian rift stage, the plate was stretched, and the blocks of the Earth's crust diverged south of the chain of trough depressions along dipping ledges, forming a Graben belt. Rifting initiated simple crust stretching deformations in the field with the horizontal orientation of the intermediate and minimum stress axes $\sigma 2$, $\sigma 3$ under the vertical axis of maximum stress $\sigma 1$. Open to crust-mantle fluids, vertical detachment fractures, and faults formed in the rocks. Parallel to the extension of the rift belt was the axis of intermediate stress ($\sigma 2$), orthogonal was the extension axis ($\sigma 3$). Rifting was controlled by shifts-transforms in the northeast diagonal strike-slips in azimuth directions 24–30°, 39–45°, and 54–63°. Along with the transformations, "tectonic rift rails" were formed, along which the rift shoulders diverged in sub-latitudinal fault zones, and the segmentation of the future rift belt was laid. The tectonic position of the Devonian rift on the mobile outskirts of the EEC contributed to active inductive lithosphere deformations under the influence of the movements of the Tethys's oceanic plates [8–10]. The rift's southeast flank was located on the northern outskirts of Tethys and opened into the area of Late Paleozoic sub-oceanic basins and island arcs. During the Late Hercynian-Early Cimmerian activation, this area was transformed into the basement of the Scythian-Turan Plate. Platform paleo-basins known as "Dnieper-Donets depression" and "Donbas" existed from the Middle Devonian to the Early Permian [8, 9]. This is evidenced by relics of the rift structure, Devonian alkaline-basaltic volcano, and folded Paleozoic platform cover. Both basins ceased their existence due to fold deformation at the Late Hercynian Saal' phase, about 0,250 Ga [24, 26].

The results of the study of stress fields of the Sarmatia [3, 8–10, 16, 26–28] indicate that tectonic inversion and folding in the southeastern segment of the DDP began at the end of the Early Perm (Saalian and Palatinate phases) of the Late Hercynian tectogenesis. It is believed that at the first stage, there were inversion deformations according to the model of oblique left-sided collision (**Figure 6**). They were influenced by the orogenic movements of the northern front of the collisional orogen formed on the active plate within the Paleotethys Ocean [8, 9]. In the body of the Sarmatian plate, this collision led to the formation of a stress field of tangential compression falling in the northeast direction.

Mesozoic and Cenozoic tectonic movements within the Sarmatia plate are considered as a consequence of orogenic processes in the Black Sea-Caucasian segment of the Paleo-Ttethys folded orogen [8–10]. These processes caused the formation of a northwest tangential stress field in the Sarmatia Earth's crust (**Figure 6**). At the neotectonic stage of geological evolution, the Paleorift structure continues to develop in geodynamic conditions of deep collisional compression and near-surface extension in the regional horizontal shear stress field [29, 30]. The mode of deep compression may have been caused by both the colliding geodynamic processes of the interaction of active plates and orogenic mobile belts on the northern edge of the Tethys paleo-ocean, as well as the intrusion at the late stage of riftogenesis of the mantle plume with uplift and structural and material processing of the base of the lithosphere [8, 9]. At the second stage of tectonic inversion, in the Mesozoic and Cenozoic the progression of the Donbas inversion uplift and its subsequent structural evolution as a DF took place in the regime of tangential compression caused by collisional stress from the Late Paleozoic Caucasus orogen [10, 27]. The progression of the general collision compression caused structural complications of Hercynian folding structure strike-slip parageneses of right-handed kinematics were formed on this territory.

Post-rift stages of the Sarmatia Earth's crust evolution are characterized by interference fields and combined deformation modes. In the collision field with horizontal axes σ1, σ2, and vertical axis σ3, vertical shear fractures are impenetrable for fluid migration and laterally penetrable detachment fractures formed through over-thrusts. Break-down structures were formed in the basin's basement by up-throws, "reverse" faults. Linear anticline folds were formed in the mantle, in orthogonal to the extension of which was the σ1 axis, and in parallel was the σ2 axis. Through the interference of collision stress modes and shifts in the field with horizontal axes σ1, σ3, and vertical axis σ2, the platform mantle underwent strike-slip and cover-thrust deformations. The geodynamic situation was characterized by a sub-meridian orientation of the stress axis with a variable inclination to the southwest (Saal' phase), north (Laramian

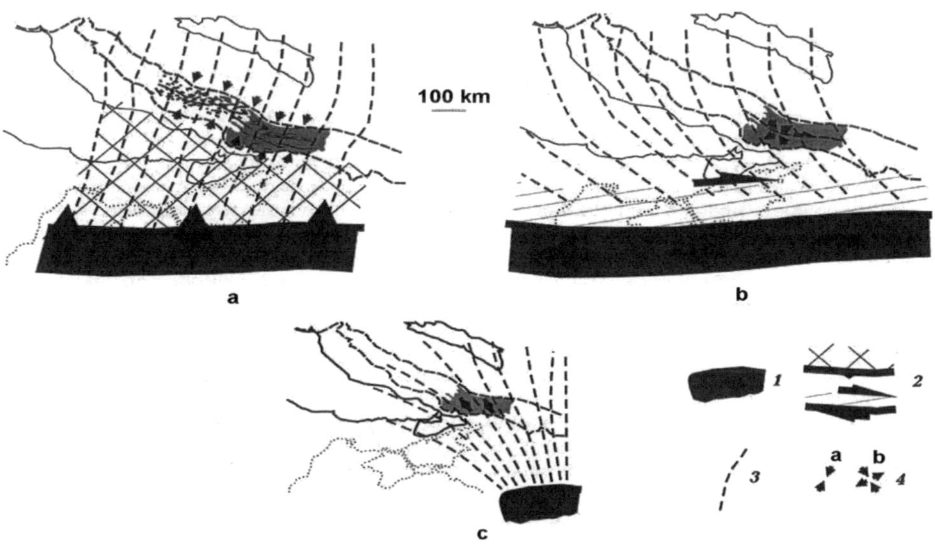

Figure 6.
Schemes of geodynamic regimes of the Sarmatia at the tectonic inversion stage of the Dnieper-Donets Paleorift, by [29]: a—collision mode of the initial stage of inversion; b—local transtension in the mode of right shift of the main stage of inversion; in—the scheme of induction influence of the remote collision orogen. Symbols: 1—mobile area of the active environment; 2—deformation conditions: top—collisions, bottom—strike-slip deformations; 3—model trajectories of compression stresses; 4—modes: a—tangential compression; b—strike-slip deformation.

phase of the Early Alpine), and northeast (Attic phase of the Late Alpine) [16, 26]. The stress field was influenced by the pressure from the Donbas, which underwent tectonic uplift. In the Cimmerian (Donets phase) and Alpine (Laramian phase) activation, this one underwent cover folded deformations and transformed into the DF [8–10, 16, 26]. In the adjoined West Donets segment, the Mesozoic complex is eroded, three structural surfaces—Hercynian and two Alpine (Laramian and Attic), the structure of which is controlled by age-differentiated deformation grids of crust-mantle sutures [16, 26], were formed in the platform cover. Hercynian folding took place in the over-thrust-strike-slip mode of left-handed kinematics, while Cimmerian and Alpine cover-over-thrust in the strike-slip field of right-handed kinematics [8–10, 27]. In the Alpine flore of the West Donets segment and on the west slopes of the DF, under the overthrust framework, expressed in the Donets Ridge cover-throw scales were formed [16]. Displacement of blocks on the southern slope of the Graben caused the superimposition of sediment-volcano rift and platform cover at several stratigraphy levels by over-thrust plates of crystalline rocks. On the northern slope of the USh, the South Donbas Melange Zone was formed, manifested in relief as the Azov Heights [16].

Thus, the evolution of the Earth's crust of the Sarmatia counts several significant stages due to the geodynamic situations and tectonic modes that arose on the EEC [2, 3, 6, 13, 16, 24, 28]:

1. Paleo-proterozoic collision of Fenno-Scandia and Sarmatia in the uplift mode with the unification of the crystalline massifs of the USh and the VA;

2. Early Riphean "athermal" rifting with the splitting of the "cold" continental crust along the linear germinal strike-slip zone, forming the chain of Trogs by the Pull-apart mechanism;

3. Devonian epi-continental "arch-thermal" rifting through the divergence of the crust under the influence of the mantle plume, forming the Graben sedimentary basin;

4. Hercynian-Early Cimmerian activation under collision conditions against the background of synchronous subsidence of the superimposed platform depression with the formation of linear folded zones involving salt tectonics;

5. Late Cimmerian-Alpine activation in the situation of interference of collisional stress and regional strike-slip field with territory uplift, basin desiccation and the formation of the West Donets Cover Folded Region and the DF.

3.3 Fault-block divisibility of Earth's crust of the Sarmatia

In the Earth's crust of the Sarmatia, eight azimuth directions of the Planetary fracture grid used by faults have been detected [2]. Faults of different directions in geochronology differ in genetic type, direction, and kinematics of displacement in geochronology: 285–290° (PR3, O, S—chips, strike-slips, ₽—over-thrusts, up-throws) 310–315°(D-C1—over-thrust, up-throws, N,—chips, strike-slips), 340–345° (PR3—gaps, throws, C1-2, C2-3, P2, T2-3—chips, strike-slips), 0° and 90° (D—chips, strike-slips) 15–20°(PR3, O, S—chips, strike-slips, ₽—gaps, throw), 45–50° (gaps, down-throw), 70–75° (PR3—over-thrust, up-throws, C1-2, C2-3, P2,

T2-3—chips, strike-slips). Based on 3D seismic exploration data, the properties of tectonic lineaments of the DDP have been determined. Due to the statistical analysis of over 30,000 faults of the basement and four structural surfaces of the sedimentary cover, six realized azimuth directions of the Planetary grid have been established: 1—273–279° and 9–18°, 2—282–288° and 24–30°, 3—291–312° and 39–45°, 4—315–339° and 54–63°, 5—342–351° and 72–78°, 6—354–6° and 84–90° (**Figure 7**) [6, 15]. There are two directions (291–312°, 315–339°) in the northwest diagonal system of faults located along the extension of the Paleorift, creating a typical strike-slip pattern for the inner structure of the deformation zones (**Figure 8**). In the pre-rift stage, they laid the germinal linear shear zone.

There are three directions (24–30°, 39–45°, and 54–63°) in the northeast diagonal system formed shifts-transforms, which controlled the process of rifting. The faults of the diagonal and orthogonal system in azimuth directions (273–279°, 291–312°,315–339°, 354–6°, 9–18°, 39–45°, 54–63°, and 84–90°) changed structural-dynamic manifestation at the stages of platform activation (**Figure 9**) [6, 13]. The "reverse systems" faults were part of the tectonic grids of folded layers of different ages in the basin cover. Changes in the dynamics and kinematics of "inverse systems" occurred in a new geodynamic setting due to space–time inversions of the stress field tensor. Inversion deformations initiated changes in the tectonic mode and rebuilt the rift structure with a new plan controlled by a younger tectonic framework.

Thus, there are three structure plans of deformation at the Phanerozoic stages of structural evolution of the Sarmatia Earth's crust were formed: the rift (D2-C1t), the platform syncline (C1v – P1), and the collision (P2-T—Q) (**Figure 9**) [6, 13, 14].

3.4 Distribution of inversion deformations in the Dnieper-Donets Paleorift

It is a priori assumed tectonic stresses and deformations during the platform activation first affected the Precambrian basement of the Sarmatia as a base layer of the Earth's crust [2, 6, 13, 14]. Later, the deformations covered the basin sedimentary cover [6, 16, 26, 29]. The deformation of the rift structure increases along to the southeast.

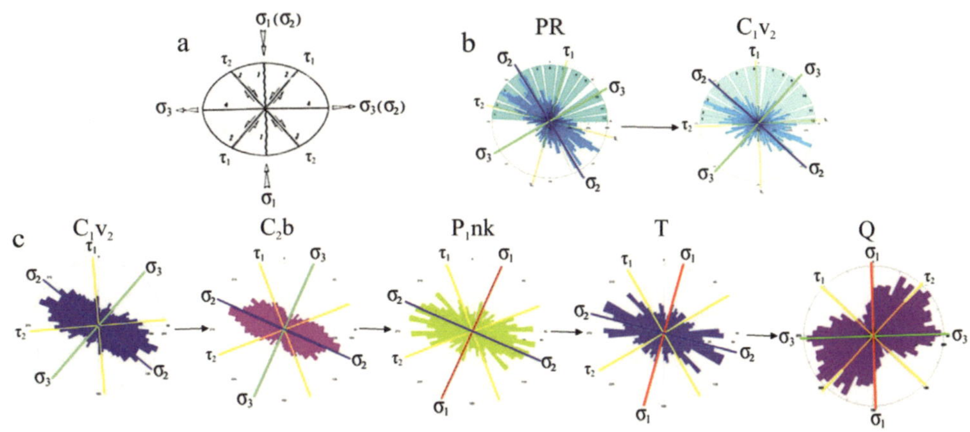

Figure 7.
Pie charts of the spatio-temporal distribution of the stress field tensor of the Sarmatia Earth's crust in the Dnipro-Donets Paleorift. A—scheme of the deformation ellipsoid tensor, by [3], B—the regional plan, C—the central part (Lokhvytsa, Izyium segments). The stress field tensor: б1—the maximum normal stresses; б2—the average normal stresses; б3—the minimum normal stresses; τ1, τ2—the paar of maximum tangential stresses.

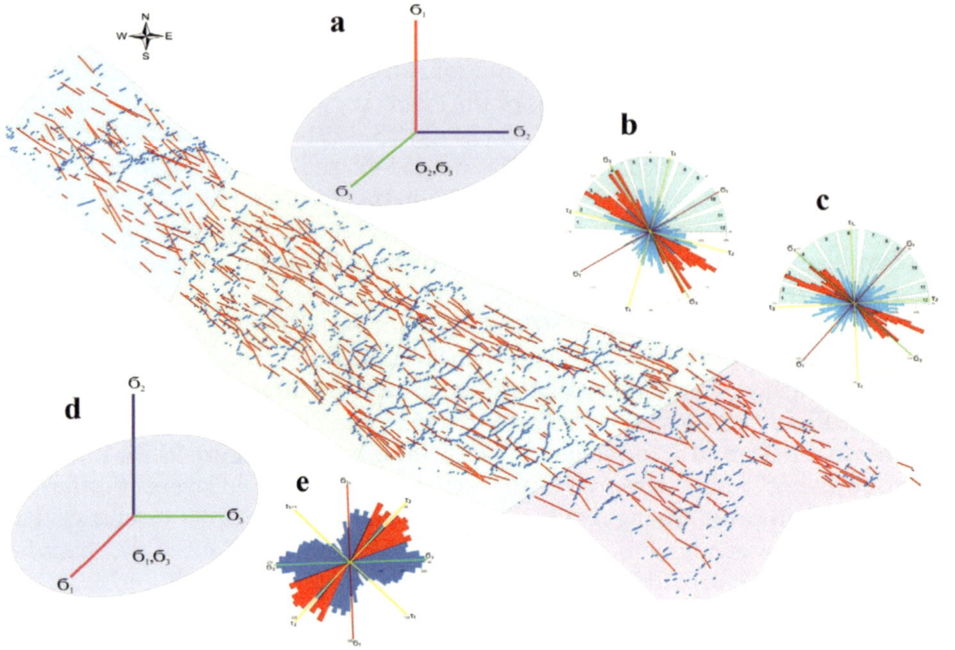

Figure 8.
Comparative tectonic scheme of rift and post-rift fault frames and stress field tensor of the Sarmatia Earth's crust in the Dnieper-Donets Paleorift in the Precambrian basement, sedimentary cover, and topography. Insets: a, d—stress field tensor: a—rift stage; d—collision stage (Attica); b, c, e—diagrams of the azimuthal distribution of geodynamic axes: b—in the foundation; c—sedimentary cover, e—in relief.

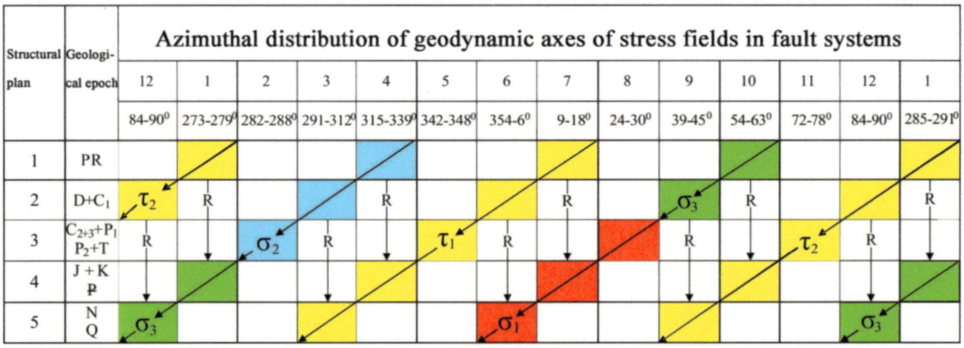

Figure 9.
Periodically scheme of the kinematic mechanism of the spatio-temporal inversion of the stress field tensor of the Sarmatia and the formation of the deformations structural plans of the Dnieper-Donets Paleorift in Phanerozoic, by [6]. Symbol: R—"reverse faults"; another symbol in ***Figure 7****.*

Rift throw relics remained in the West micro-plate, and on the northern flank of the East micro-plate along sharp changes in basement depth, reflected by elongated gravity field gradient anomalies. The structural boundary between weakly dislocated

(Chernihiv, Lokhvytsia, Izium) and completely inverted (West Donets, DF) segments is drawn along the Central-Azov-Slovyansk crust-mantle suture at the meridian of Balaklia city (**Figure 10**) [29]. Here, the north (Kochubiyevka-Volvenkovo) and south (Sosnivka-Stepkovka) chains of axial salt dome anticlines are interrupted (**Figure 11**). To the east of the border extends the only axial Druzhkivka-Komyshuvakha salt wall. The rift framework and Hercynian folded lines were segmented by diagonal strike-slips into coulisse sections. Border throws bend in the plan initially to the east, and at the meridian of Donetsk city to the southeast, as a result of which the area of the Graben expands in the West Donets segment. Due to Hercynian movements, the rift structure underwent an uplift of several km and linear folding. In the north of the West Donets, the vector of Hercynian movements ascends to the southwest and crosses at a large angle the eroded surface of the basement. This contributed to the over-thrusting of Precambrian basement rocks onto the Paleozoic cover along the gentle North Donets overthrust (**Figures 12** and **13**). In the south Graben border with the Azov Massif of the USh, the basement surface serves as a plane of tectonic ruptures due to a northeast azimuth of fall coinciding with the Hercynian movement direction (**Figures 6** and **14**). By the same kinematic mechanism, the South Donbas Melange Zone in the Paleozoic complex has been formed. The Melange Zone did not develop further west due to a structural barrier—a ledge of basement under the southern slope of the Graben (**Figures 13–15**). The west slope of the ledge is bypassed by an ensemble of small horst and grabens of rift stretch. Their axes turn from south to southeast, rise up, and cut at sharp angles at tectonic contact with the Zone of Melange. The rift stretch structure is orthogonally superimposed by the frame of uni-directional "unconformity" down-throws. Because of this, the hanging blocks at the base of the grabens lie 2–3 km higher than the foot blocks. The rise in the pre-Triassic erosion level to the southwest indicates pre-Mesozoic denudation from the southern slope of the Graben to the Azov massif with a vertical amplitude of more than 3 km.

The West Donets Cover Folded Region was formed in the Hercynian and Alpine folded floors in the West-Donets segment and on the west slopes of the DF (**Figure 11**). The

Figure 10.
The map of the pre-rift faults of the Dnieper-Donets Paleorift Precambrian basement, by [31]. Symbols: 1—the sedimentary cover local structures; 2—salt doms; 3–5—fields: 3—gas; 4—gas and condensate; 5—oil; 6–11—faults: 6—of the borders, 7—of the basement, of the Devonian, of the Carboniferous, of the Permian, of the local structures; 12–13—the changes of the sedimentary complexes thickness: 12—by stratigraphic distribution, 13—by vertical amplitudes and dip; 14—boundary zones of fold deformations; 15—boundaries of the West-Donets Cover Folded Region.

Figure 11.
Fragment of the tectonic map of the Dnieper-Donets Basin, by [31]. Symbols: 1—the Precambrian basement surface marks, m; 2—seismic profiles: a—old, b—new, c—reinterpreted; 3—deep wells: #/mark; 4—faults: a—throw, b—thrusts, c-d—diagonal strike-slips of extension: c—north-west, d—north-east; 5—boundaries of the West-Donets Cover Folded Region; 6—anti- and sin-structures; 7—Devonian salt domes; 8—boundaries of the tectonic zones; 9—the basin boundaries; 10—compensatory throughs, 11—depressions, 12—anticlines, 13—anticlines shaft, 14—salt-dome anticlines zones: 1—Kochubiyvka-Volvenkovo, 2—Sosnivka-Stepkivka, 3—Druzhkivka-Komishuvakha, 15—ledges.

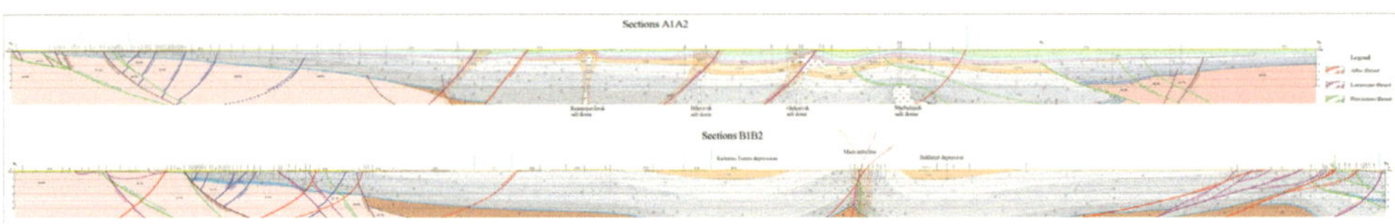

Figure 12.
Regional geologic sections of the Dnieper-Donets Basin, by [30]. A1A2—through the south-eastern part of the Izium segment of the Dnieper-Donets Basin. B1B2—through the central part of the West Donets segment.

wedge-shaped region between Luhansk, Izium, and Donetsk cities consists of echelons of scales of tectonic covers and coulisse-jointed anticlines. The multi-age strike-slip and thrust frame controls the "cross-thrust" structure of deformations (**Figures 12** and **14**). The north flank of the Region is bounded by the North-Donets thrust, in the south the Samara and Novoselyvka thrusts separate it from the Mélange zone. The Region is divided in half into two structural areas: (1) Luhansk-Komyshuvakha Thrust Folding Area, (2) Kalmius-Torets Thrust Scally Area by the axial Sulino-Kostyantynivka fault. In the axial zone of the Graben, echelons of over-thrusts control tectonic blocks of elongated east and northeast stretch, complicating the structure of the Bakhmut, Komyshuvakha-Liman, and Kalmius-Torets depressions. In the front of the blocks, thrust anticlines were formed with steep northeast wings and gentle southwest ones, which turn into monoclines in the rear. On the west flank, the sole of the Mesozoic cover plunges to the west to a depth of over 3 km within the Orchik depression in the Izium segment.

The Cover Folded Region is located above a triangular depression in the basement, filled with a sedimentary thickness of up to 20–22 km (**Figure 10**). In the center of the Region, in the zone of the Axial (Sulin-Kostyantynivka) upthrow-shear, the largest Main (Horlivka) anticline is located, with a length of 170 km at a

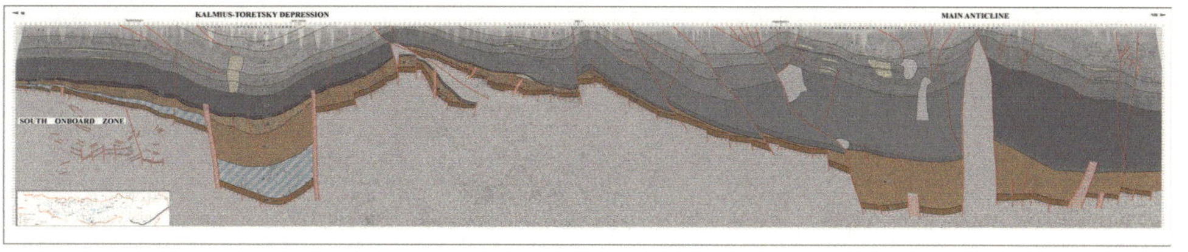

Figure 13.
*Pokrovs'k-Biryukovo seismic section through the south slope of the West Donets Graben and the axial part of Donets Foldbelt, by [29]. The inset is the profile line on the map (**Figure 11**).*

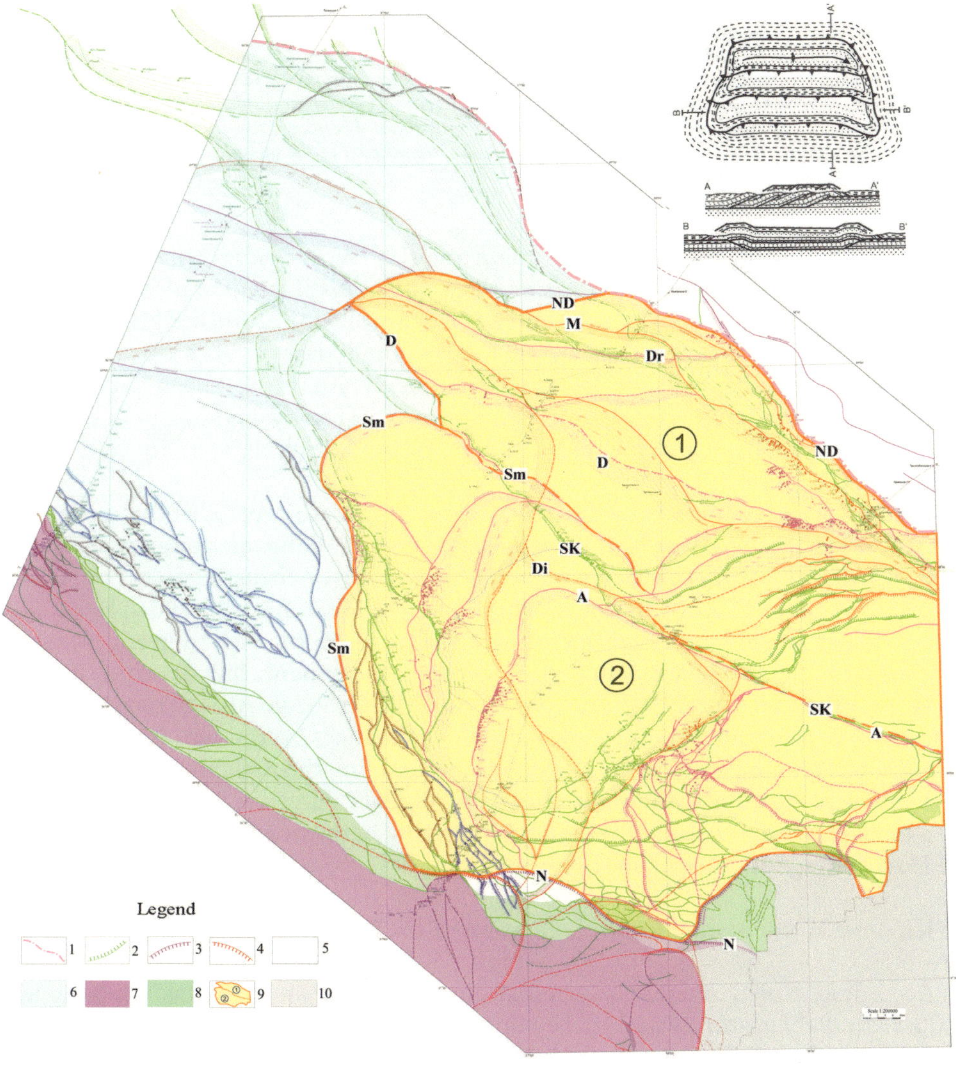

Figure 14.
Scheme of the West-Donets Cover Folded Region tectonic frame. Symbols: 1—north border of the region; 2–4—thrusts: 2—Herzinian, 3—Laramian; 4—Attic; 5—low-dislocated Paleozoic autochthon; 6—Herzinian neo-autochthon; 7—the Priazov Massif; 8—the South Donbas Melange Zone; 9—West-Donets Cover Folded Region: (1) Luhansk-Komyshuvakha thrust folding area, (2) Kalmius-Torets Scally of Thrust Area; 10—the Donets Foldbelt. Insert—scheme of the cover folded system.

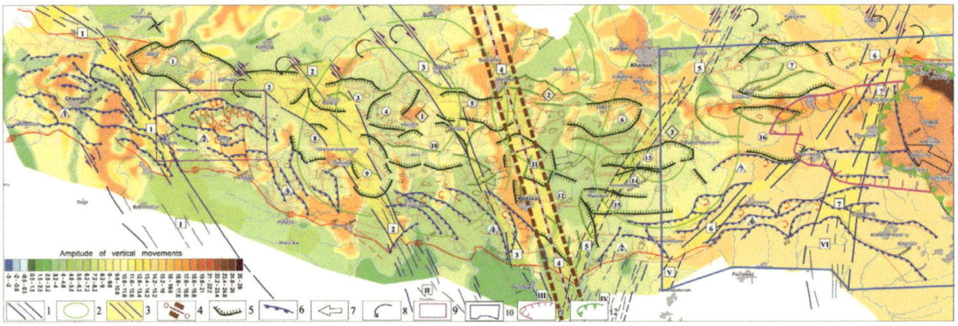

Figure 15.
Kinematic model of the Dnieper-Donets Paleorift inversion deformation of Earth's crust on the map of vertical neo-tectonic (Alpine, Holocene) movements. Symbols: 1—pre-rift deep faults; 2—ring anomalies on satellite images; 3—neo-tectonics strike-slip zones, 4—the Verkhovtsy-Lgov tectonic axis of kinematic symmetry; 5–6—structural parageneses of deformations: 5—tension, 6—compression; 7–8—rocks displacement directions; 9—the frontal part of the Nizhyn-Ichnia structural paragenesis of longitudinal bending; 10—the most dislocated area of the basin; 11—West-Donets Cover Folded Region; 12—the frontal part of the indenter horst mega-block of the Donetsk Foldbelt. Another symbol in ***Figure 1****.*

width of 10 km (**Figures 11–14**). The main anticline with a ridge hinge was formed in the mode of lateral over-compression through the displacement of wings along the left-lateral up-throw and strike-slip with an amplitude of tens of kilometers. In the southeast direction, the hinge of the fold gradually rises by 8–9 km, as a result of which Mesozoic deposits on the wings are eroded. On the northern flank of the orocline, in the hanging wings of Hercynian thrusts (Drobyshivka, North Donets, and Matross), coulisse-sectioned chains of upthrow-anticlines were formed (**Figures 11**, **12**, and **15**).

The Late Alpine (Oligocene-Miocene) deformations are controlled by an over-thrust framework, behind which large blocks of the pre-Cenozoic complex are segmented into plates from the first to dozens of km and over-thrusting in the northeast direction (**Figures 12** and **15**). The maximum amplitude of horizontal displacements of rock blocks reaches 4.5 km with a vertical one up to 1.5 km. Attic deformations are only absent in the southwest part of the West Donets segment, they are significant in the rest of the territory. To the north and east, the Cenozoic surface rises sharply over 250 m toward the Donets Ridge. Altitude differences exceed hundreds of meters, corresponding to the amplitudes of displacements along the Attic thrusts. Attic movements are manifested in right-lateral movements along the Samara, Novoselivka over-thrusts in the south and the North Donets one in the north, and the formation of the Donets Ridge and the Azov Highlands.

3.5 Kinematic mechanisms of deformation of the Earth's crust

According to instrumental determinations in the DDP and its outskirts, basement faults are mostly up-throw and strike-slips [3, 16, 26–28, 31]. In the sedimentary cover, in the shear field above them, rows of diagonal strike-slips were formed in correspondence to the coulisse arrangement of fracturing chips [5, 13, 29]. Due to the combination of elementary deformation modes (compression/expansion/displacement) in the fault zones, parageneses of "flower" and "duplex" structures are formed [32]. In the trans-tension mode, "tulip" and "contractional duplex" structures are formed in paragenesis of upthrow, strike-slipes, and overthrusts. In the extension mode, "palm tree" and "extensional duplex" structures are formed in paragenesis of throws, strike-slipes. Lateral displacements of rocks are controlled by synthetic and antithetic strike-slips, which are composed of R and R'-fracturing [7, 32]. Structure dislocation is formed along shears both co- and counter to the movement direction in the wings of faults and due to their combination, with a rotation and without it [7].

During the Hercynian collision, the rock displacement within the rift basin occurred toward the "geodynamic shadow" zones [29]. Evidence of longitudinal extrusion of rocks is the stress measurements on volcanic structures in the articulation zone of the FD with the Azov Heights [28]. In trachytes, two fields of shear type were found: Late Alpine (Attic) with a sublatitudinal compression axis and a submeridional extension axis, and early Alpine (Laramian) with the reverse location of the stress axes. In andesites, diagonal zones of slating of northwest and northeast extensions are established. In separate bodies, a late Hercynian strike-slip field of right kinematics is formed. A geophysical feature of movements along faults is stock-like zones of loss of correlation of seismic reflective horizons in the wave field due to tectonic fragmentation and loosening of rocks in the fault zones [5]. Evidence is a large stock structure in the seismic field at the hinge of the main anticline (**Figure 13**).

Hercynian movements caused the displacement of the masses of rocks from over-compressed areas in the central part of the rift to the largest by area and less compressed West Donets segment. The discharge area of the extruded masses widens in a wedge-shaped manner along the course of the tectonic flow from the extrusion area at the eastern border of the Izium segment to the western slopes of the Donbas (**Figures 11** and **14**). The orocline outlines of the unloaded massif coincide with a triangular depression in the basement filled with 20–22 km of rocks. Above the depression, there is an ensemble of arcuate bent, coulisse-linked, and strongly compressed upthrow-anticlines, transverse in extension to the direction of flow. Transverse folding is diagonally superimposed on the lines of large Hercynian anticlines of rift orientation (**Figure 11**). The apparent reverse overthrust of upthrow-anticlines backward along the flow indicates a lack of geological space at the point of its origin. Due to the situation of volume compression, the upper layers of allochthon lagged behind the general movement of the flow, therefore a structure of underthrust under transverse folds was formed here, not vice versa. Strong stress in this area initiated a high orogenic uplift of layers, the formation of secondary echelon overthrust-folding, and at least three intersecting structural levels (Hercynian and two Alpine). The masses extruded with the flow accumulated here synchronously with loading. Evidence of this is the intersection of the fault-thrust frame and zones of near-fault folding in the west—at the starting point of the flow origin (**Figure 14**). Further, along the flow, the excess of masses gradually dissipated due to stress reduction. Therefore, transverse thrust structures arise, and longitudinal thrust-folds are replaced by transverse ones. As the orocline expands, large folds elongate and turn into anticlinal swells. This is evidence of their formation due to tectonic disruption and thrusting onto a paleo-depression on the western slope of the DF.

Alpine movements were directed northwest toward the central part of the inverted belt. The lateral stress was induced by the movements of the mountain-folded Caucasus and the inversion uplift of the Donbas, which turned into a folded belt. Alpinotype deformations occurred due to the movement of masses from the west slopes of the DF along the previously formed orocline framework (**Figure 14**). Under the pressure from the east of the megablock-indenter of Donbas, the masses flowed in the opposite direction to the Hercynian current (**Figure 15**). As a result of their accumulation on the Hercynian neo-autochthon, two alpine (Laramian and Attic) systems of thrust covers were formed (**Figures 12–14**). This situation led to a significant thickening of the sedimentary cover in the literally limited space of the paleo-basin due to the thrusting of younger allochthon plates (**Figures 14** and **15**). The allochthon composed a post-sedimentation part of folded surfaces of the cover that was not compensated by sedimentation. According to the kinematic mechanism of transverse protrusion along dynamically conjugated and kinematic antithetic diagonal up-throw-slips without a rotational component of movements, a "cross-over thrust" structure of the "orocline of transverse protrusion" was formed (**Figure 16**) [7, 29]. The Cover Folded Region and the Melange Zone were formed on the south slope of the Donbas. The Melange Zone is expressed in the relief as the Azov Highlands stretches diagonally to the Donets Ridge and has similar elevation marks.

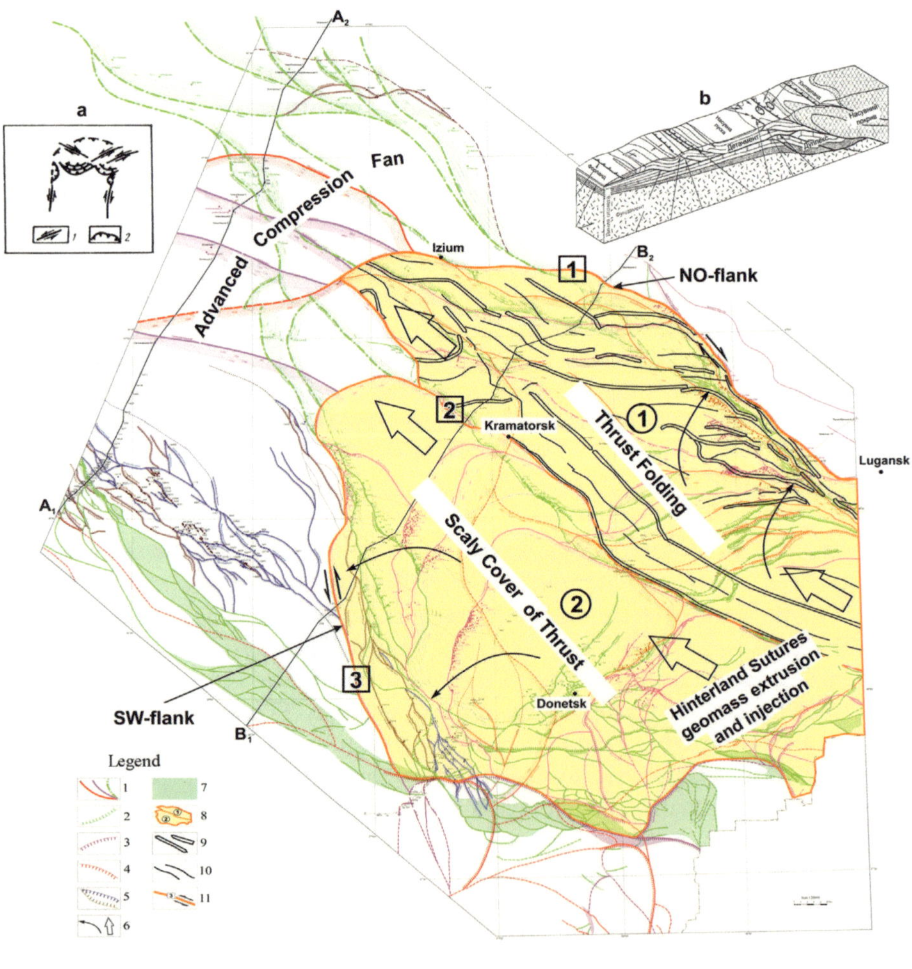

Figure 16.
Kinematic model of formation of the West Donets Cover Folded Region. Symbols: 1—advanced compression fan of the tectonic oroclin; 2–4—thrusts: 2—Hercynian, 3—Laramian; 4—Attic; 5—rift relics throw; 6—rocks displacement directions; 7—South Donbas Melange Zone; 8—West-Donets Cover Folded Region: (1) Luhansk-Komyshuvakha Thrust Folding Area, (2) Kalmius-Torets Thrust Scally Area; 9—anticlines: 10—sinclines; 11—linear strike-slip zones: [1]—north-east flank, [2]—Axial, [3]—south-west flank and kinematic of displacements. Inset: a—scheme of the "Longitudinal Extrusion Orocline"; b—scheme of the cover folded system.

The sublatitudinal rock displacement along the fault frame caused deformations of the longitudinal elongation of the rift structure due to the stretching of the rock layers along their strike [4, 7]. According to the first kinematic mechanism, the structure is divided into several tectonic blocks by rows of antithetical shifts of the same kinematics. The blocks jointly rotate like a row of dominoes, along shifts diagonal to the axis of the structure (**Figure 17**). Shear displacements are directed against the movement of the general flow of rock masses. The ensemble of dominoe blocks experiences unidirectional movements orthogonal to the stretching of their curved wings, which leads to an elongation of the original structure along its length [7]. In the West microplate on the southern slope of the Graben, a behind-the-scenes ensemble was identified, consisting of the Chernigov, Nezhin-Ichnya, and Priluky-Lubny dynamically related structures of longitudinal extension (**Figure 18**).

According to another mechanism, longitudinal tension is realized through multidirectional motions of blocks with rotation along two dynamically coupled diagonal strike-slips [7]. Due to the tectonic spreading of the rocks in opposite directions, a new intermediate trapezoidal block is formed in the fault zone (**Figure 19**). Subsequently, the rocks are squeezed out of the block, which experiences tensile deformation. Tension zones are formed in the basement, depressions are formed in the sedimentary cover above them. Such a geodynamic situation developed in the central part of the Graben at the boundary of the West and East micro-plates in the zone of dynamic influence of the Verkhovtsy-Lgov crustal-mantle fault (**Figure 15**). As a result of displacements of rocks in its wings, diagonal strike-slips like as a plumage were formed. Due to multidirectional movements with rotation along diagonal

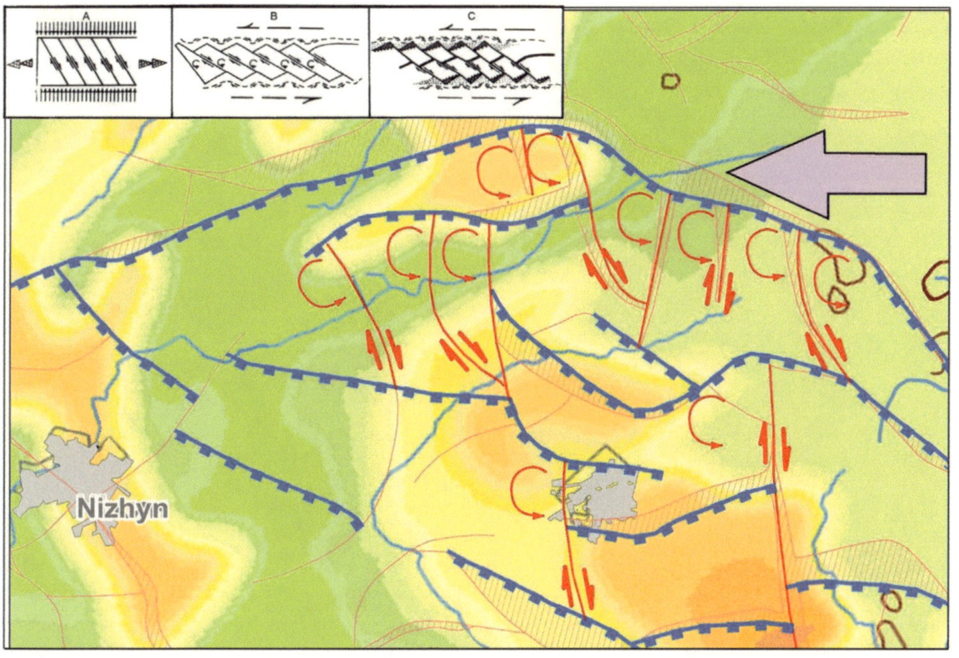

Figure 17.
The internal kinematics of the frontal part of the Nizhyn-Ichnia parageneses of the longitudinal bending in the central part of the Paleorift on the map of Holocene vertical movements. The block "dominoes structure" was formed due to the longitudinal displacement with the rotation of the blocks and on diagonal shifts of the same kinematics. Inset: A kinematic scheme of a block's "dominoes structure," by [7]; A, B—before and after deformation; C—a structural pattern.

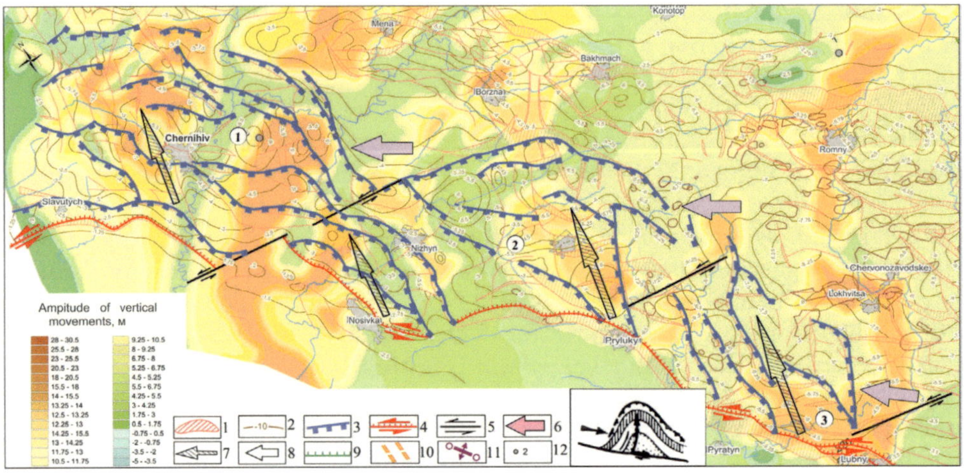

Figure 18.
*Model of a longitudinal bend kinematic mechanism of rift structure deformation in the south part of the Paleorift on the map of Holocene vertical movements. A number of the oroclines of longitudinal bending: ①—Chernihiv; ②—Nizhyn-Ichnia; ③—Pryluky-Lubny. Inset: Kinematic scheme, by [7]. Symbols (**Figures 17** and **18**): 1—basement faults; 2—basement surface marks, m; 3—strike-slip-throw frame of the oroclines; 4—edge shear-fault; 5—shear-transforms; 6—rocks displacement directions; 7—directions of extrusion and shear flow; 8—vectors of local tension; 9—arcuate fault-spreading; 10—the Verkhovtsy-L'gov deep fault; 11—tectonic axis of kinematic symmetry; 12—wells.*

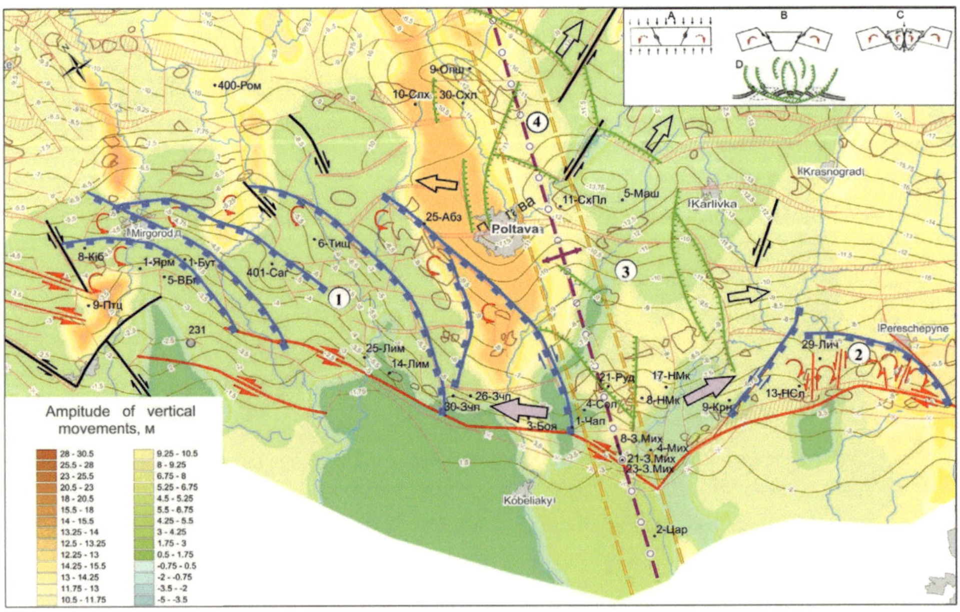

Figure 19.
*Model of a "Transverse Spreading Fan" kinematic mechanism of rift structure deformation in the central part of the Paleorift on the map of Holocene vertical movements. Inset: A, B, C—before and after deformation; D—a structural pattern. Number of the structural paragenesises: ①—the Mirhorod orocline of longitudinal bending; ②—the Pereshchepino orocline of longitudinal bending; ③—the Vorskla lenticular pull-apart basin; ④—the Verkhovtsy-L'gov deep fault. Symbols in **Figure 18**.*

strike-slips, intermediate tectonic blocks were formed. The situation of divergence of the Earth's crust initiated the spreading of crystalline rocks in opposite directions outside the blocks. Tensile deformations initiated the formation of an ensemble of local lenticular depressions in the basement surface, elongated along the meridian. Above them, in the sedimentary cover, the Vorskla depression was formed by the Pull-apart mechanism (**Figure 20**) [30]. The geodynamic region of the Earth's crust extension is controlled by the strike-slip paragenesis of the tectonic "longitudinal extension fan" turned to the north (**Figure 21**) [33].

At the collision mode, synchronous deformations of lateral tension and vertical extension also occurred [7]. In the basin confined space, under the influence of lateral stress, there was a tectonic extrusion of rocks upward free from compression of the ground surface. The uplift of masses on the surface initiated the formation of stress zones and uplifts over the fault zones as a source of erosion and wear in the basin. Examples are the Olenivka, Volodarka, Uspenka, and Rovenets over-thrust anticlines, which are reflected in the heights of the West Donbas relief (**Figure 13**). At the same time, a compensatory geodynamic factor caused the formation of extension zones of the Earth's crust, dynamically and spatially conjugated with stress zones [2, 3, 15, 27]. As a result, "overlaid" depressions were formed, where the products of erosion accumulated, removed from nearby hills. Examples are from the largest Kalmius-Torets, Bakhmut, to the smallest Mykhailivka, Kuybyshevka, and other depressions are reflected in the basin's relief (**Figures 12** and **13**). Modern uplifts and associated troughs are located diagonally to the strike of basement fault zones due to horizontal displacements of rock masses in the limbs of strike-slips in the platform cover (**Figure 11**). Together they form a segmented paragenesis of longitudinal elongation.

Thus, at the inversion stage, the rift basin underwent uplift and folding deformations, transforming the rift and post-rift platform complexes into folded structural floors of the cover. The tectonic framework of deformations formed by remobilized

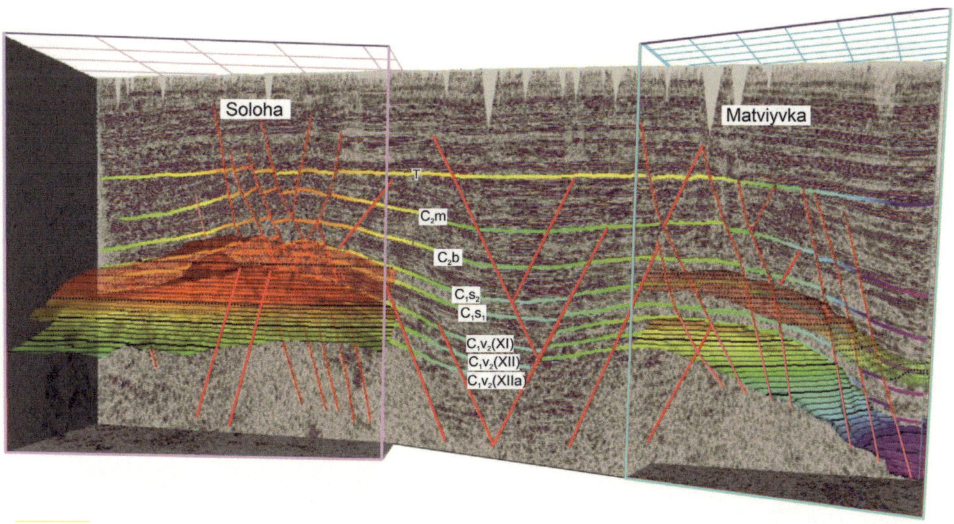

Figure 20.
Model of the "transverse spreading fan" kinematic mechanism on a 3D tectonic reconstruction at the end of the Late Hercynian Saal' phase in a section through the Vorskla depression in the central part of the Paleorift.

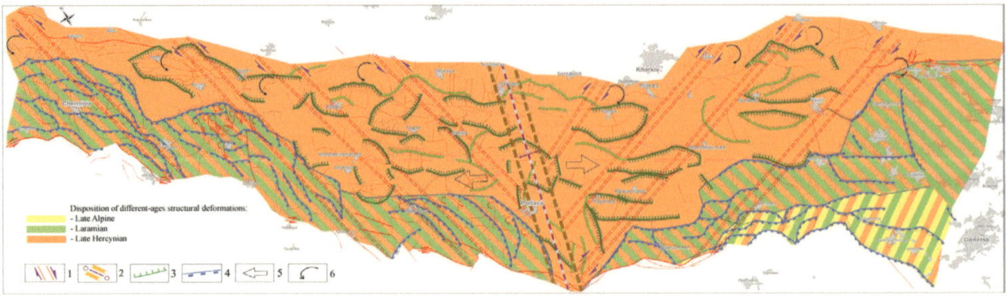

Figure 21.
Kinematic scheme of inversion complicated of Dnieper-Donets Paleorift Earth's crust: 1—regional linear shear zones; 2—the Verkhovtsy-L'gov tectonic axis of kinematic symmetry; 3–4—folded longitudinal zones: 3—arcuate fault-spreading, 4—strike-slip-throw frame; 5–6—direction of blocks displacement and rotation.

pre-rift crustal-mantle faults is superimposed diagonally and orthogonally on the network of rift throws. The strike-slip and uplift network controls the regional inversion structures of "longitudinal extrusion orocline of rocks" in the southeastern part and "the transverse extension fan" in the central part of the DDP. Therefore, paleobasins "Dnieper-Donetsk Depression" and "Donbass" of the Paleozoic complex at the inversion stage of the evolution transformed into the West Donetsk Cover Folded Region and DF due to Hercynian and Early Alpine deformations. Currently, the inversion structure of the Paleorift is overlain by Mesozoic and Cenozoic complexes of the platform cover and additionally complicated by Alpine dislocations.

4. Conclusions

1. The stress field of Sarmatia was influenced by inductive movements of the Arabia and Scythia, spreading of the North Atlantic Ridge, pressure from the Caucasus and the indentor mega-block of the FD. Anisotropy of the Sarmatia is manifested from surface to base of the lithosphere in variations of composition and thickness of layers of consolidated crust, the Moho, the lithosphere, and the crust base. The structural-material differentiation of the lithosphere determined the processes of evolution of the Sarmatia Earth's crust, including the division into the West and East micro-plates, the formation of a rift structure, and the inversion deformations of the DDP. A defining factor of the evolution of Earth's crust is the remobilization of faults during platform activation and the formation of a deformation framework.

2. The deformation framework was formed along the pre-rift crust-mantle faults, diagonally superimposed on the rift fault structure frame. There are three plans of deformation were formed: rift (D2–C1t), platform syneclise (C1v–P1), and collision (P2–Q) in the Paleorift. The intensity of deformations increases along the rift strike to the southeast. So, in the West segments relics of the rift are preserved, but in the East ones, the structure is destroyed by Alpine deformations. Deformations are manifested in the formation of tectonic detachments, upthrows in the basement; coulisse strike-slip deformation framework of the sedimentary cover; "cross-overlapping over-thrust" structure of cover folded deformations of the Hercynian and Alpine; thickening of the cover in the limited space of the basin due to the over-thrusting of younger tectonic allochthons that make up the uncompensated by sedimentation part of the cover. The regional

collision framework of deformations is formed by the parageneses "orocline of longitudinal extrusion of rocks" in the southeastern part and "fan of transverse spreading of rocks" in the central part of the Paleorift.

3. The paleo-basins "Dnieper-Donets Depression" and "Donbas" by Hercynian and Cimmerian deformations of the Paleozoic complex turned into the West Donets Cover Folded Region and Donets Foldbelt, were buried under the Mesozoic and Cenozoic platform cover and were further complicated through by Alpine deformations. Considering the ubiquitous distribution of inversion deformations, a conclusion is made about the transformation of the DDP into a Dnieper-Donetsk Movable Foldbelt. Neo-tectonic deformations are structurally manifested by the Donetsk Ridge and the Azov Highlands. Considering the ubiquitous distribution of inversion deformations, a conclusion is the transformation of the DDP into a Dnieper-Donets Movable Foldbelt.

Acknowledgements

I thank V. Karazin Kharkiv National University, Ukrainian Research Institute of Natural Gasses, Geological Enterprises of Ukraine for providing materials. The author is grateful to S.V. Goriaynov, V.I. Alekhin, V.A. Korchemagin, O.B. Gintov for sharing their measurements of stress fields in Ukraine, S.V. Goryainov for providing his sections and the structural map of the West Donbas Hercinides, V.I. Starostenko for the schemes of the Sarmatia lithosphere, Polivtsev A.V. for the map of vertical neo-tectonic movements.

Author details

Oleksii Bartashchuk
Kharkiv Karazin National University, Kharkiv, Ukraine

*Address all correspondence to: alekseybart@gmail.com

References

[1] Wilson JT. Did the Atlantic close and then re-open? Nature. 1966;**211**:676-681

[2] Chebanenko II. Theoretical Aspects of Tectonic Divisibility of the Earth's Crust. Kiyv: Naukova dumka; 1977. p. 84 (in Russian)

[3] Gintov OB. Field Tectonophysics and Its Application in the Study of Deformations of the Earth's Crust of Ukraine. Kyiv: Phoenix; 2005. p. 572 (in Russian)

[4] Chikov BM. Introduction to the physical foundations of static and dynamic geotectonics. In: Kontorovich AE, editor. Sib. Dep. of the Academy of Sciences, V. Sobolev Institute of Geology and Mineralogy. Novosibirsk: Publishing house "Geo"; 2011. p. 299 (in Russian)

[5] Timurziev AI. Experience in the use of tectonophysical methods to increase the efficiency of prospecting, exploration and development of the near-strike-slip oil. Geophysical Journal. 2014;**2**(36):172-185 (in Russian)

[6] Bartashchuk OV. Evolution of the stress-deformed field of the Earth crust of Dnieper-Donets paleorift at Phanerozoic. Reports of the National Academy of Sciences of Ukraine. 2019;**3**:62-71 (in Ukraine)

[7] Kopp ML. The problem of space for deformations arising in the shear stress field (on the example of the Mediterranean-Himalayan orogenic belt). In: Strike-Slip Tectonic Disturbances and their Role in the Formation of Minerals. Moscow: Nauka; 1991. pp. 75-85 (in Russian)

[8] Stampfli GM, Borel GD. A plate tectonic model for the Paleozoic and Mezozoic constrained by dynamic plate boundaries and restored synthetic ocean izochrons. Earth and Planetary Science Letters. 2002;**196**(1-2):17-33. DOI: 10.1016/S0012-821X(01)00588-X

[9] Natal'in BA, Sengor AMC. Late Paleozoic to Triassic evolution of the Turan and Scythian platforms: The prehistory Paleo-Thethian closure. Tectonophysics. 2005;**404**(3-4):175-202. DOI: 10.1016/j.tecto.2005.04.011

[10] Meijers MJ, Hamers MF, van Hinsbergen, et al. New late Paleozoic paleopoles from the Donbas Foldbelt (Ukraine): Implications for the Pangea A vs. B controversy. Earth and Planetary Science Letters. 2010;**297**(1-2):18-33. DOI: 10.1016/j.epsl.05.028

[11] Bogdanova SV, Pashkevich IK, Gorbatschev R, Orlyuk MI. Riphean rifting and major Palaeoproterozoic crustal boundaries in the basement of the East European Craton: Geology and geophysics. Tectonophysics. 1996;**268**(1-4):1-21

[12] Starostenko VI, Rusakov OM, Pashkevich IK, et al. Tectonics and Hydrocarbon Potential of the Crystalline Basement of the Dnieper-Donets Basin. Kiev: Galaxy; 2015. p. 212 (in Russian)

[13] Starostenko VI, Pashkevich IK, Makarenko IB, et al. Heterogeneity of the lithosphere of the Dnieper-Donets depression and its geodynamic consequences. Part 1. Deep structure. Geodynamics Series. 2017;**1**(22):125-138 (in Russian)

[14] Granovsky AG. Magmatism and stages of the tectonic evolution in the Donets folded structure, eastern Donets Basin. Geotectonics. 2018;**4**:96-110. DOI: 10.1134/S0016853X18040033

[15] Bartaschuk OV. System organization of disjunctive tectonics of consolidated basement in Dnipro-Donets paleorift. Part 1. Lineaments. V. Karazin Kharkiv National University, Series Geology. Geography. Ecology. 2016;**48**:12-27 (in Ukrainian)

[16] Goryainov SV. Cenozoic tectonic movements of Eastern Europe. Geology and Minerals of the World Ocean. 2013;**3**:55-65 (in Russian)

[17] Heidbach O et al. The world stress map database release 2016: Crustal stress pattern across scales. Tectonophysics. 2018;**744**(2):484-498

[18] Zoback ML. First and second order patterns of stress in the lithosphere: The world stress map project. Journal of Geophysical Research. 1992;**97**:11703-11728

[19] Becker A. Detached neotectonic stress field in the northern Jura Mountains, Switzerland. Geologische Rundschau. 1989;**78**(2):459-472

[20] Bott MH. Ridge push and associated plate interior stress in normal and hot spots regions. Tectonophysics. 1991;**200**(1):17-32

[21] Cloetingh S, Wortel R. Stress in the Indo-Australian plate. Tectonophysics. 1986;**132**:49-67

[22] Xie FR, Chen QC, Cui XF, et al. Fundamental database of crustal stress environment in continental China. Progress in Geophysics. 2007;**22**:131-136

[23] Atmaoui N, Kukowski N, Stockhert B, Konig D. Initiation and development of pull-apart basins with Riedel shear mechanism: Insights from scaled clay experiments. International Journal of Earth Sciences. 2006;**95**:225-238

[24] Sydorenko G, Stephenson R, Yegorova T, et al. Geological structure of the northern part of the eastern Black Sea from regional seismic reflection data including the DOBRE-2 CD profile. In: Sosson M, Stephenson R, Adamia SA, editors. Tectonic Evolution of the Eastern Black Sea and Caucasus. Vol. 428. Geological Society, London, Special Publications; 2017. pp. 307-321

[25] Wüstefeld A, Bokelmann G, Barruol G. Evidence for ancient lithospheric deformation in the East European Craton based on mantle seismic anisotropy and crustal magnetics. Tectonophysics. 2010;**481**(1-4, 15):16-28

[26] Korchemagin VA, Emets VS. Peculiarities of the development of the tectonic structure and stress field of Donbas and the eastern Azov region. Geotectonics. 1987;**3**:49-55 (in Russian)

[27] Kopp M, Korchemagin V. Cenozoic stress/strain fields of the Donbas and their probable sources. Geodynamics Series. 2010;**1**(9):17-48 (in Russian)

[28] Alekhin VI, Korchemagin VA. Stress and strain fields of subvolcanous bodies and dikes Permian-Triassic age of the Azov block of the USh and its articulations with the Donbas. In: Gliko AO, Leonov JG, editors. Tectonophysics and Topical Issues of Earth Sciences. Abstracts of the All-Russian Conference. Vol. 1; Schmidt Institute of Physics of the Earth, Moscow. 2008. pp. 197-200. Available from: https://www.researchgate.net/publication/298793117 (in Russian)

[29] Bartashchuk OV, Suyarko VG. Geodynamics of formation of the transition zone between the Dnieper-Donets basin and the Donbas foldbelt. Tectonic regimes and kinematic mechanisms of inversion. Geodynamics Series. 2021;**1**(30):25-35. DOI: 10.23939/jgd2021.01.025

[30] Bartashchuk OV, Suyarko VG. Horizontal displacements of geomassages in continental rifting geostructures (on the example of the Dnieper-Donets paleorift). Part 2. Structural paragenesis of shear deformation of the sedimentary cover. V. N. Karazin Kharkiv National University. Series Geology. Geography. Ecology. 2019;**50**:27-41 (in Ukrainian)

[31] Bartashchuk OV, Suyarko VG, Chuenko OV. Tectonic inversion of the Dnipro-Donets basin. Part 3. Tectonic style of deformations. Visnyk of V. N. Karazin Kharkiv National University. Series Geology. Geography. Ecology. 2023;**58**:29-37. DOI: 10.26565/2410-7360-2023-58-01 (in Ukrainian)

[32] Sylvester AG. Strike-slip faults. Geological Society of America Bulletin. 1988;**100**:1666-1703

[33] Bartashchuk OV, Suyarko VG. Geodynamics of formation of the transition zone between the Dnieper-Donets basin and the Donbas foldbelt. Tectonic inversion of rift-like structure. Geodynamics Series. 2021;**2**(31):53-65. DOI: 10.23939/jgd2021.02.053

Chapter 3

New Insights into Geodynamic Evolution of the South-Eastern Termination of the Tunisian Atlas during Early Cretaceous Period from Surface and Subsurface Data: Hydrogeological Implications

Mohamed Ben Chelbi

Abstract

The Menzel Habib Plain (MHP), the easternmost part of the Southern Tunisian Atlas, exhibits a thick siliciclastic and carbonate succession developed in the Early Cretaceous period. Integration of seismic and gravimetric data coupled with analysis of the syndepositional faults affecting these units proves that the MHP is governed, during this period, by a N-S to NE-SW extensive tectonics induced by the sinistral drifting of Africa with respect to Europe and the beginning of opening of the Eastern Mediterranean Sea. The geodynamic evolution of the MHP is mainly due to the jerky normal activities of the N-S, NW-SE, NE-SW and E-W oriented faults. During extensive period these network of faults architects the sedimentary floor into succession of horst and grabens inducing formation of E-W to NW-SE basins. Seismic lines show that the actual architecture of the MHP obtained after Tertiary tectonic inversion, is that of succession of large synclines separated by narrow anticlines. Porous and permeable unites forming these deep structures host very developed aquifers characterizing the sub soil of MHP and were totally different to the surrounding aquifers of the southern Tunisian margin.

Keywords: Menzel Habib plain, hydrogeology, extensive tectonics, continental drift, deep aquifers

1. Introduction

The Tunisian Atlas, an eastward extension of the Maghrebide chain (**Figure 1A**) [1–5], had long recorded a polyphase tectono-sedimentary evolution of the North African margin [6–9]. Since the initial breakup of the Pangea to the Tethys and Mesogean rifting periods [10, 11] and through the collision events that continue [12, 13], the basins evolution recorded tectono-sedimentary systems with complex facies and structures.

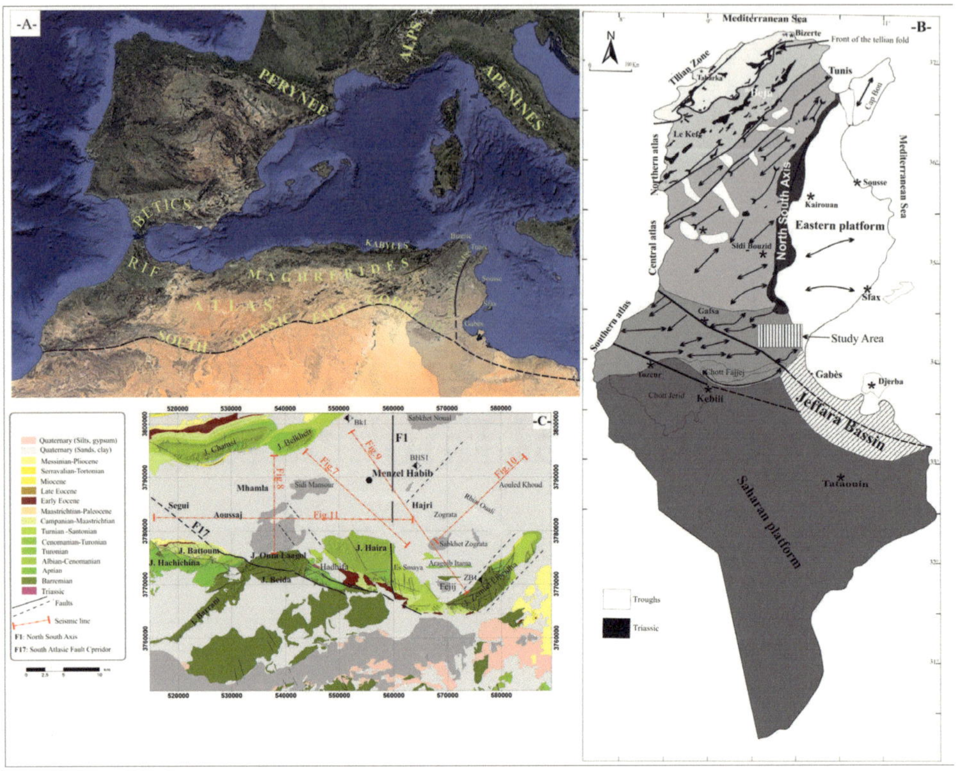

Figure 1.
Geological context of the study area. A- general position of the Tunisian margin in the Mediterranean context; B- structural zoning of Tunisia [6] showing the structural position of the study area in relation with the major structures of the central and southern atlas, the eastern platform and the Tunisian Jeffara. C- geological map of the study area showing the most master faults (F1 and F17) which affect the Menzel Habib plain (MHP) and the position of the various seismic lines used in this study.

Geological work that has targeted the Tunisian margin has shown that this complexity is mainly due to a direct interaction between the brittle tectonics, materialized by the different network of faults, and the sedimentation [7, 14–20]. Menzel Habib plain (MHP), constituting the Southeastern termination of the Tunisian Atlas (**Figure 1B**), was not spared from this tectono-sedimentary interaction. This plain characterizes a range of stratigraphic and structural interference between all the major structures of the Tunisian margin (**Figure 1B**). It occupies an umbilical position between the Eastern platform, the Central Tunisian Atlas, the Southern Atlas, and the Jeffara (**Figure 1B**). Geodynamically, the evolution of this zone is closely controlled by the jerky activities of the North-South Axis (NSA) and the South Atlasic Fault Corridor (SAFC) represented by the Hadhifa Fault (**Figure 1C**). By its complex structural and stratigraphic position, the MHP shows a capital interest, both geodynamically and hydrogeologically, to understanding the larger neighboring structures evolve at the expense of major faults that the shreds. Although the geological position of the Menzel Habib plain is very important, works that have targeted this area are very rare and have only focused the mapping of the various stratigraphic and structural units [21]. The aim of this work is (1) determining the different fault networks that affect the study area; (2) showing the tectono-sedimentary evolution of the study area in relation to the major fault systems based on field and subsurface data, (3) giving the outline of the hydrogeological characteristics of potentially aquifer formations, and (4) correlating the evolution of this zone with other examples in neighboring regions.

2. Geological framework

2.1 Stratigraphy

(**Figure 1C** and **2**). Triassic beds crop out in Jebel Hadhifa, composed of clay, marl, limestone, gypsum, salt, anhydrite, and sand. The Triassic Germanic facies, typical of the Tunisian margin, is evident. The Lower Cretaceous is outcropped only by the top part of the Bou Hedma formation, the Sidi Aich, and Orbata ones. The BHS, ZB1, BK1, and MAN1 oil wells characterizing the MHP and surroundings (**Figure 1C**) served to study the non-outcropped formations. In the Chott Fejej area, Mejri et al. [22] attribute the Early Cretaceous successions to the Sened and Gafsa Group defining the Meknassy Super Group [23]. The Melloussi formation, 200 m, comprises interbedded sandstones and claystone with minor dolomite and anhydrite. This formation has been deposited on a shallow marine shelf with an irregular subsidence. The Boudinar formation is made up of white sand, fine to coarse-grained, with fine interbedded levels of green shale, and occasionally thin levels of brown dolomite. The average thick of this formation is 180 m. These two formations constitute the Sened group largely described in the surrounding areas of Fejej Chott.

The Gafsa group, starts with the Bou Hedma formation essentially made up by claystone with interbedded sandstone, dolomite, limestone gypsum, and occasional coal. 100 to 150 m of fine white sands characterize the summit of this formation, which usually ends with 3 m of red dolomite forming a boundary between the Bou Hedma formation and the Sidi Aich one. At the core of Zemlet Elbeidha anticline, the Bouhedma formation is represented by 1250 of gypsum, claystone, sandstone, and dolomite alternations (**Figure 2**) in which 800 m are crossed by the ZB1 petroleum well. The Sidi Aich formation is sandy unit widely spread in Central and Southern Tunisia. In our study area, this formation is composed of interbedded sand and varicolored claystone with occasional dolomite and rare traces of coal. The average thick of the Sidi Aich formation turn around 80 m. This last is usually overlapped by 40 to 60 m of brown and massive dolomites attributed to the Orbata formation. At the South-Eastern flank of the Zemlet Elbeidha anticline, these dolomites are replaced by a bar of polygenic conglomerates formed by dolomite blocks embedded in a sandy and clayey matrix. These dolomites are capped with another dolomitic cliff constituting the base of the Zebbag formation, with a major unconformity. Consequently, if the base of Orbata is approximately synchronous, the top is completely diachronous depending on the sedimentation of the upper members, the subsequent erosion, and the onlaps of the following series [22].

Upper Cretaceous stratigraphyincludesthe Albian beds, which is represented by the Zebbag formation (Upper Albian to Cenomanian), the Aleg formation (Turonian to Lower Campanian), and the Abiod formation (Lower Campanian to Upper Maastrichtian) (**Figure 2**). These formations are composed of dolomitic, gypsum, and clay alternations, in the lower part, argillaceous and marly deposits, in the middle part, and dolomites and limestones, in the upper part.

The Tertiary and Quaternary consists of predominantly unconsolidated sand and conglomerates at the base overlain by red clays with crystalline gypsum. These deposits are very rare and generally occupy the piedmonts of the great reliefs.

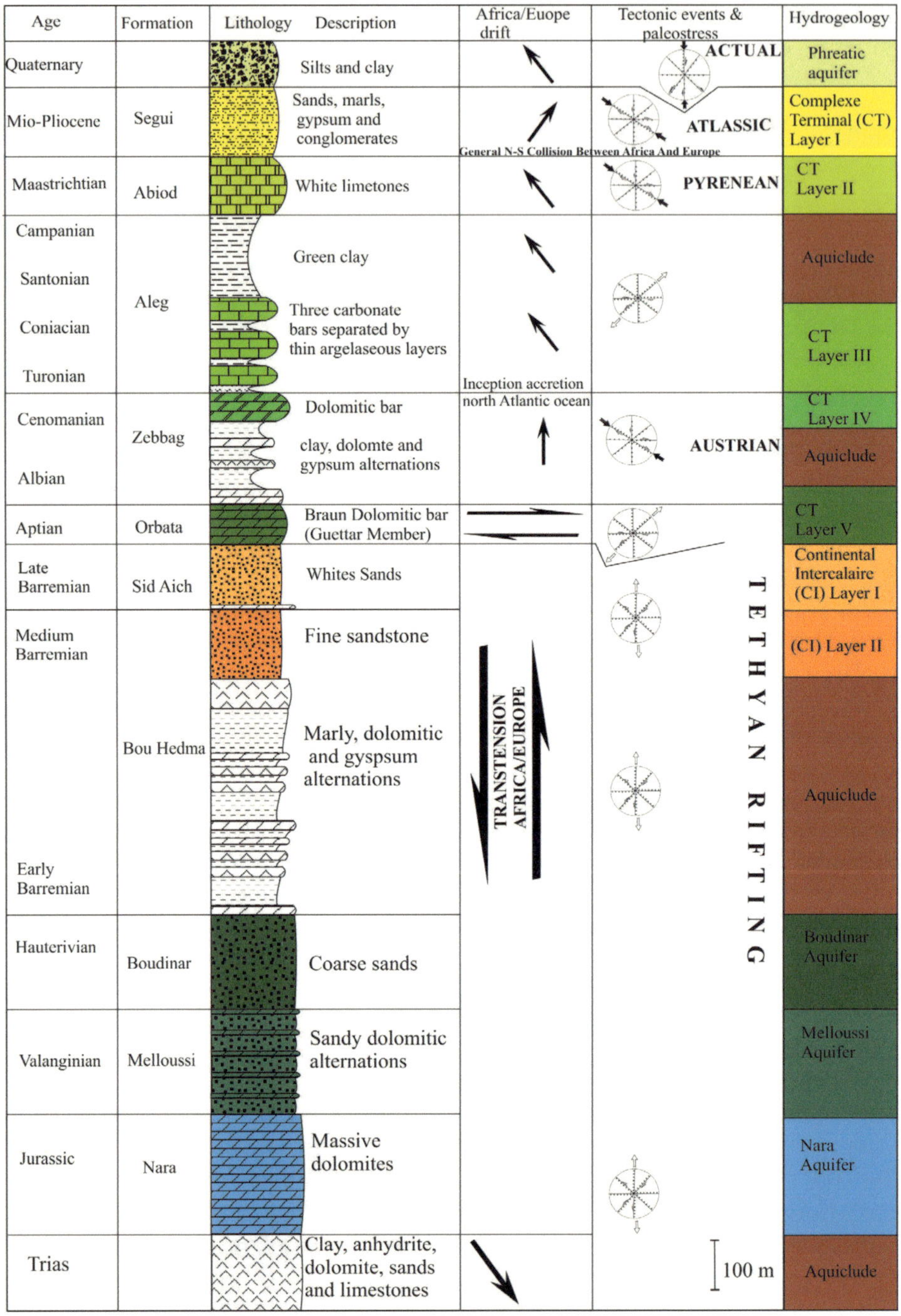

Figure 2.
Geodynamic and stratigraphic context of the study area; column 1, 2 and 3 show the different lithostratigraphic unites outcropped or crossed by the various oil wells in the MHP; column 4 and 5 expose the principal stages of Africa drifting with respect to Europe and the corresponding tectonic events recorded in the Tunisian margin since Triassic (Ben Chelbi 2013); column 6 summarizes the corresponding hydrostratigraphic units characterizing the subsoil of MHP.

Palaeogeographically, deposits of the Melloussi and Boudinar formations reflect a deltaic regressive detritic body evolving to a deltaic progradation toward the north [23]. Deposits of Bouhedma, Sidi Aich and Orbata formations characterize transgressive sedimentation on a low slope substratum. Facies and thickness changes of these formations are due to the activities of the great faults acting during extensive epochs contemporaneous with sedimentation and to the jerky halokinetic rises of Triassic salt bodies.

2.2 Hydrostratigraphy

Despite the fact that almost all of the MHP is occupied by Plio-Quaternary deposits (**Figure 1C**), the subsoil contains all the geological formations known for their qualities of forming good aquifers of regional scope. Based on the petrographic and structural characteristics of these different geological formations, as well as on previous work undertaken in the area and its surroundings [21, 24–27], we have identified and defined several units that may correspond to potential aquifers (**Figure 2**). Some of these are well known and well-studied, others are not yet and are targets in other neighboring basins.

The most superficial aquifer is lodged in the sandy, silty, and clay deposits of the Pio-Quaternary. In some places, these series are very thick, reaching up to a hundred meters. For decades and considering these hydrogeological characteristics and their accessibility, the Plio-Quaternary formations are extensively exploited. We count more than 350 surface wells that have been dug in these series [21]. Most of these wells are dry, and the rest have a very high salinity of around 5 to 10 g/l.

The second aquifer is known as the "Complexe Terminal" (CT) (**Figure 2**), which characterizes the Southern Tunisia and South-Eastern of Algeria. This aquifer is a multilayered system stored the Miocene sandy formation, the Upper Cretaceous carbonate members (Campanian-Maastrichtian), and the Zebbag dolomitic formation. Unusually, the Aleg formation, which outcrops at Jebels Belkheir, Chamsi, Berda, and Zemlet Elbeidha is represented, at its base, by fractured white limestones. For this, the latter can be considered as intermediate terms characterizing the TC. In the studied area, only a deep well, which has reached these carbonates and given water with a low salinity (2 g/l). The massive and fractured dolomites attributed to the Orbata formation could constitute an additional aquifer, which can be added to that of the CT due to the stratigraphic continuity and which exists between the two terms.

The Barremian sands, defining the Sidi Aich formation and the summit part of the Bouhedma one, omnipresent in the basements of the MHP, store the most important deep aquifer characterizing the South of Tunisia. This aquifer is called "Continental Intercalaire" (CI) (**Figure 2**). These aquifers mentioned above are well-known and have been exploited for a long time either for agriculture or for industry.

Considering their petrographic and structural characteristics, the Boudinar and Melloussi formations, respectively represented by sands and dolomitic, and sand alternations may constitute the deepest aquifers of the MHP. These formations were crossed by the BHS1 petroleum well and have shown that they store appreciable quantities of water. Because of their enormous depth, these two last groundwater levels are note previously studied and defined.

Hydrogeological studies undertaken in this area [24, 25] have shown that these aforementioned aquifers are recharged from large reliefs in the North, West, and South, respectively represented by Jebels Belkheir-Chamsi, Jebel Berda, and the

North chain of Chotts (Jebels Bir Oum Ali, Jebel Hachichina, Jebel Batoum, Jebel Oum Laagol, Jebel Haira-Es Smaya, and Jebel Zemlet Elbeidha). Moreover, these aquifers are recharged by vertical infiltrations from the endoreic basins (Sabkhet Sidi Mansour-Mhamla, Garaet Hajri, Sabkhet Fejij, Garaet Wali and Garaet Aouled Khoud) through a very dense network of active faults.

2.3 Structural evolution

Since the Triassic period, the Tunisian margin has recorded all the echoes of the geodynamic evolution of the Tethyan domain and the resulting structures (**Figure 2**) [4]. Indeed, from the first stage of the breakup of Pangea in the early Triassic, a system of regional NS, NE-SW, and NW-SE trending faults greatly jagged sedimentary floor into a mosaic of a variables blocks and expanses. From this period a rifting system was implemented, applying these faults, in response to an extensive tectonics involving a N-S minimum stress. Evidence of the first stages of the Tethyan rifting is dated to the Jurassic-Barremian period [28–31]. This mechanism of opening is continuing until Barremian period (**Figure 2**). A radical change occurs during the Aptian that materializes by changing the direction of elongation to become NE-SW [6, 32]. This change in the minimum stress orientation is indicated by reactivation of the NW–SE-trending faults with normal activity, which created several NE–SW and NW-SE basins [6, 15]. These will evolve during the following periods and give the different troughs of the central Tunisian Atlas [15] and the major escarpment of the Southern Tunisian Atlas and Jeffara Zone [1]. Contemporaneous with this activity, the Tunisian Atlas records intense halokinetic activity (**Figure 2**) forming a large system of diapirs and/or salt glaciers [14, 33].

Tunisian margin was submitted, during the Albian Cenomanian period, to compressive tectonic applying NW-SE maximum stress (**Figure 2**) [7]. This dynamic activates the N-S, NE-SW, and NW-SE existing faults into sinistral reverse, reverse, and normal dextral activities, respectively.

During the Late Cretaceous the Tunisian margin was controlled by transtensional to transpressional tectonics [6, 9], characterized by reactivation of the north–south-, east– west-, and NE–SW-trending normal faults with dextral component. The general inversion of the structures was recorded during the middle Eocene [14, 15, 19, 32] (**Figure 2**). The first period of folding is assigned to the Tortonian [34], whereas the ultimate compression leading to the building of the Tunisian Atlas Mountains began during the Villafranchian and continues to the present day (**Figure 2**) [35]. This long period of compression controlled by NNW–SSE to N-S stress has contributed to the final configuration of the Tunisian Atlas in which NE–SW-, NW–SE-, E-W, and N-S oriented faults were respectively reactivated with sinistral reverse, dextral reverse, reverse, and sinistral slip. The Southern Tunisian Atlas has not escaped this complex evolution. The actual architecture of this part of the Tunisian margin is characterizes by coexisting of E-W trending folds, which are truncated by NW-SE and N-S oriented major faults. These last represent the signature of the SAFC and NSA, respectively. These faults have largely controlled sedimentation, during extensive period, and final structural design, during compressive epochs, of the Southern Tunisian Atlas and the Jeffara escarpment.

3. Geomorphological and hydrologic setting

Cartographically, the study area corresponds to a E-W oriented plain moderately culminating to 150 m (**Figure 3**). It is bounded on the North and South by two large

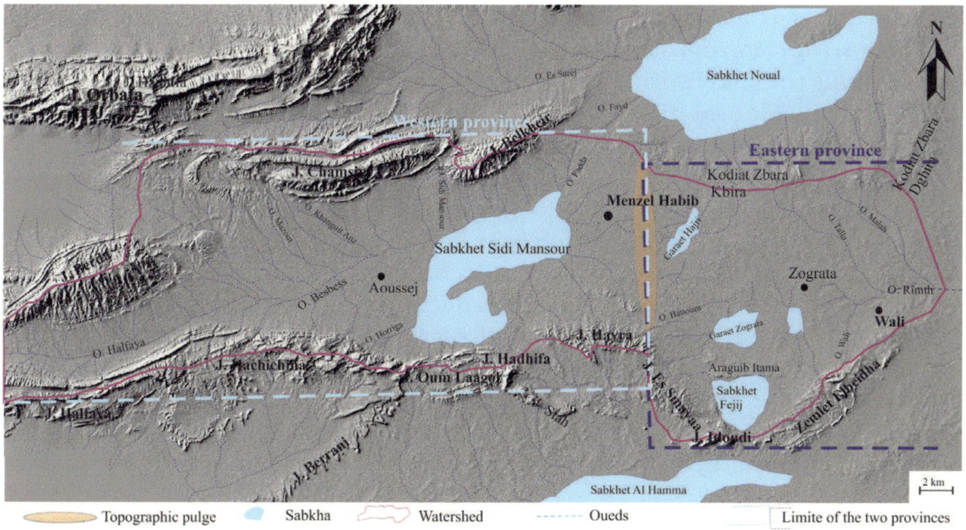

Figure 3.
Morphologic and hydrographic context of the study area.

orographic lines corresponding to the Jebel Chamsi-Jebel Belkhir and the Northern Chain of Chotts, respectively (**Figure 3**). Jebel Berda constitutes the western limit of this plain. The MHP is drained by several wadis, which flow in the Sabkeht Sidi Mansour, to the West, and in the Sabkhet Zograta and Sabkhet Ouled Khoud, to the East. The medium part of the MHP is shaped by two large topographic convexities, lengthened according to a N-S and NE-SW direction. From West to East, the first erected band is traced between Sabkhet Sidi Mansour and Garaet Hajri, while the second one lengthened the Bled et Teurbia (**Figure 3**). These convexities constitute a dividing line of discharge of the various wadis. The first band corresponds to the lay-out of a N-S major fault, which passing by Jebel Es Smaya, to the South, and continues until Jebel Belkhir, to the North (**Figure 1C**). The second convexity line represent the surface response of a deep NE-SW fault. This late is attested by the rectilinearity of the source line of the wadis, which diversify in Sabkhet Ouali. Oueds characterizing the western province flow either north in Sabkhet Noual or converge toward the Sebkhet Mhamla and Oued Rimth.

4. Data and methodology

The study of the deep structural architecture of the MHP and neighboring geo-logical structures is inferred from interpretation of gravimetric data and seismic sections, while the evolutionary geodynamic model of this zone is concluded from lithostratigraphic correlation carried out from petroleum wells and field investiga-tions. Seismic sections are obtained from the Tunisian petroleum company (ETAP). The W1 petroleum well is used to show the succession of different stratigraphic sequences characterizing the trough using time-depth conversion of lithological data. Several seismic lines are interpreted to establish the main oriented faults, which par-ticipated, firstly, to the facies repartition during sedimentation periods, and secondly to the final architecting of the study area during compressive times. Gravimetric data were the subject of a study carried out by the Ministry of Agriculture during 2016. The gravimetric campaign comprises 461 measurements distributed over a regular mesh covering the study area.

The geodynamic evolution of the MHP is interpreted, first, through the analysis of vertical facies and thicknesses variations of the Lower Cretaceous series on both sides of the NW–SE, E-W, NE-SW, and N-S oriented faults, and secondly through analysis of the main syndepositional faults affecting these formations (**Figure 1C**). Since the earliest Cretaceous, sedimentation has been largely controlled by this network of faults in which we note a large variation of facies and thicknesses. Four petroleum wells are used to show the succession of stratigraphic sequences characterizing the plain using time–depth conversion of lithological data and to correlate the lower Cretaceous formations on both sides of the major faults characterizing the study area. To understand the hydrogeological characteristics of this area, we have developed piezometric maps of deep aquifers made from well data and from depths and thicknesses maps of Lower Cretaceous formations.

5. Results

5.1 Gravity

5.1.1 Bouguer anomaly map

The Bouguer anomaly grid was established by adopting a process of interpolation of the minimum curvature. The Bouguer anomaly map expresses an amplitude of around 36 mGal, with values distributed between −42 mGal and − 6 mGal (**Figure 4**). Gravimetric responses are characterized by long wavelength anomalies arranged in a decreasing gravimetric gradient from east to west. Excess masses, reflecting the signature of positive density contrasts in relation to the density used in the Bouguer model, are organized in the eastern part of the study area (**Figure 4**). A second zone characterized by excess mass is also highlighted in the SW limit of the study zone. This subcircular geometry anomaly has a lower amplitude than those recorded in the eastern part of the study area. The western part has a mass discrepancy with increasingly negative density contrasts toward the western (**Figure 4A**). The recorded gravity gradient coincides with previous regional scale work carried out in Tunisia (**Figure 4B**). Thus, the gravimetric work of Midassi [36] reveals decreasing values of Bouguer anomaly toward the West (**Figure 4B** and **C**). We also note that this trend can be explained by the geometry of the Moho discontinuity determined by the deep seismic carried out as part of the Geotraverse program (EGT project, 1990), which expresses a lithospheric deepening toward the Gafsa region.

5.1.2 Residual anomaly and contact modeling

The residual component was calculated by subtracting the low-frequency anomalies from the total Bouguer anomaly. The map obtained offers a better characterization of the heterogeneities of local densities in the subsoil compared to the density of the Bouguer model. The residual component has more emphasis on gravimetric responses basins following short wavelength characteristics directions. The residual anomaly map (**Figure 5A**) exposes that a clear N-S dividing line differentiates the study area into two different large gravity provinces. At the western one, three

DOI: http://dx.doi.org/10.5772/intechopen.112209

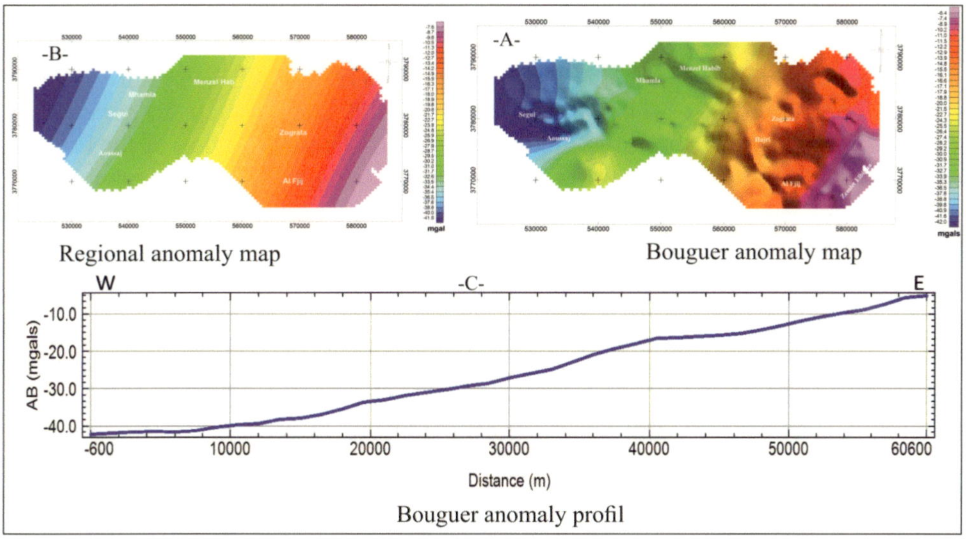

Figure 4.
A- complete Bouguer gravity anomaly map of the Menzel Habib plain; B- regional gravity anomaly map; C- E-W Bouguer anomaly profile of the study area.

basins accumulating thick series of weak density, characterizing the subsoil of Hajri, Mhamla, and Segui, are identified. The negative anomaly typifies these E-W to ENE-WSW trending zones reaching a value of −3,8 mGal. Moreover, the eastern province is made by a dense substratum reflecting a positive anomaly. The Jebel Oum Laagol, Zemlet Elbeidha, Es Smayaa, Oued Zitoun, and Rabiaat Ouali-Menzel Habib areas reaching a value 3,8 of mGal surround the Sabkhet Fejij area, which shows a negative anomaly reaching a value of −4 mGal. The passage from the eastern province with dense materials toward the western one is front and appears as a straight line defining the layout of the N-S major fault (**Figure 5A**).

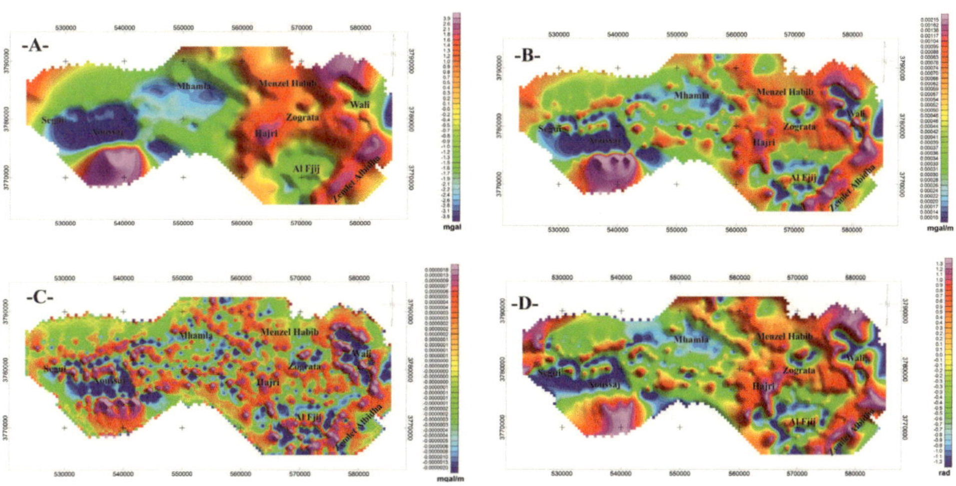

Figure 5.
A- residual anomaly map of MHP; B and C- vertical gradient anomaly maps; D- tilt angle drift map.

5.1.3 Structures and lineaments detection

To better understand the basins distribution, as well as the series which compose them, we proceeded to the elaboration of the vertical derivative maps by accentuations of the gravimetric responses (**Figure 5B** and **C**). The application of the first and second-order vertical gradient is responsible for the accentuation of geological structures near the surface (**Figure 5B** and **C**). These treatments also make it possible to better separate the anomalies and to better specify the limits of the sources. A reduction in the complexity of gravity anomalies is clearly revealed through better localization of the extensions and geometries of the basins in the study area. The basins already revealed by the residual anomaly map are well expressed by the vertical gradient anomaly maps (**Figure 5B** and **C**). The gravimetric signature of these basins shows a strong negative gradient, which proves that these accumulation zones are occupied by low-density sedimentary series. This response reflects the effects of a thick sand and clay series, attributed to the Mio-Plio-Quaternary time and the Lower Cretaceous period, which is characterized by a negative density contrast compared to the average density adopted by the model of Bouguer.

In order to determine the major directions of the different accumulation zones, already defined, we proceeded to the tilt angle derivative technics. This last is widely used to define the limits of gravity sources (geological structures of interest located on the subsurface). Thus, the limits of the sources are expressed by a zero angle. The tilt angle drift map reveals strong gravity gradients occupying the limits of the identified basins (**Figure 5D**). The geometries released regain that revealed by the first and second-order vertical gradients maps.

The horizontal gradient technics is generally used to highlight lateral variations in density, which may correspond to faults. This technic also makes it possible to characterize the major structural directions of the geological contacts. The latter are represented by gravimetric alignments, which represent the maximum amplitude on the vertical gradient map. The interpretation of the maxima of the horizontal derivative, thus made it possible to identify the discontinuities of density in the subsurface. These discontinuities are interpreted as the possible signatures of major fault systems across the region. The horizontal gradient map (**Figure 6A**) clearly shows that the study area is differentiated into several subareas, each characterized by various structures and faults. A major N-S fault divides the study area into two provinces. That of the East is characterized by NE-SW, E-W, and NW-SE the western province is affected by faults essentially oriented E-W.

The structural map obtained after application of these different technics shows several lineaments (**Figure 6B**), which highly contributed in the geodynamic

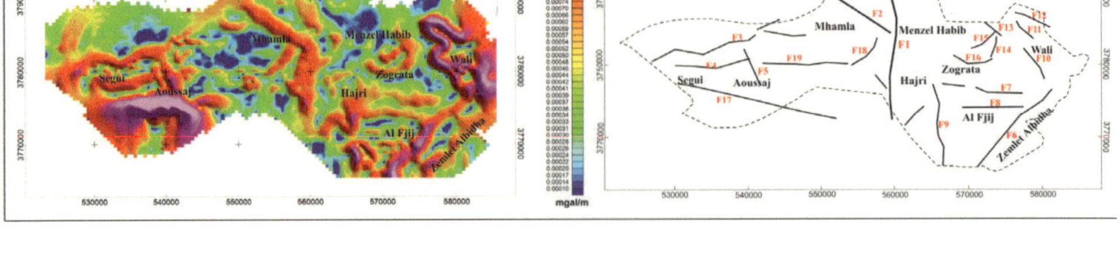

Figure 6.
A- horizontal gradient map of the residual gravity anomaly data; B- structural map deduced from compilation of data presented by the previously gravity mas.

evolution and the configuration of the MHP. In fact, the identified basins are ordered in two major directions. E-W basins characterize the western province of the study area, and NE-SW basins are identified in the neighborhood of Fejij and Zemlet Elbeidha.

The highlighted faults are distributed in the basement of Menzel Habib according to their amplitude and direction (**Figure 6B**). The N-S fault, dividing the study area into two clear provinces and tracing west of Jebel Es Smayaa and passing through Oued Zitoun, is the most important one. This fault, called F1 (**Figure 6B**) constitutes the prolongation, toward the South, of the mega N-S accident known as the North-South Axis. The Hadhifa fault, NW-SE direction, named F17 on the structural map (**Figure 6B**), passing north of Jebel Oum Laagol, shows a major importance in the deep structuring and the geodynamic evolution of MHP; it constitutes a segment of the SAFC.

The E-W faults are concentrated and essentially characterize the western province of the study area (**Figure 6B**). The most important fault is that which divides this province into two subbasins called Mhamla and the Segui basins. The NE-SW and NW-SE faults characterize only the eastern province of the study area. This province is also controlled by N-S and E-W faults, the most important of which are those which shred the Fejij sub-basin. **Figure 6**, which shows the main faults that have been identified and affect the basement of Menzel Habib, demonstrates that the eastern province is intensely fractured compared to the western one.

5.2 Seismic interpretation

To understand the deep architecture of the study area and highlight the role of the various faults in the geodynamic evolution of this area, several seismic sections have been interpreted. These sections, obtained from the Tunisian Petroleum Company (ETAP), are calibrated by the W1 well. The different seismic profiles used billow are transversal to the average direction of gravity discontinuities already identified previously. We present bellow five interpreted seismic lines showing the main results obtained after the dots of the different seismic horizons.

Section S 1 is oriented NW-SE (**Figure 1C**); it is traced between the two blocks mentioned above in the gravimetric study, which are separated by the N-S master fault noted F1 (**Figure 7**). The mapping of the different horizons demonstrates that this fault has largely controlled sedimentation during the Lower Cretaceous period. Indeed, the SE block exhibits a reduced series of the sandy formations of Boudinar, Bouhedma, and Sidi Aich, while they are thicker in the NW block. The NW end of the seismic section shows that the thickness of the Sidi Aich formation increases considerably. Moreover, this extremity highlights a Triassic saliferous intumescence very well pronounced at depth. The NW block, characterizing the subsoil of Mhamla, Belkhir, Segui, and Aoussaj, located between the fault F1 and the Triassic intumescence, shows that the deep architecture is dominated by a flat-bottomed synclinal in which governs the clay, sandy, gypsum deposits of the Bouhedma formation. The SE block corresponds to the Fejij plain in which the Bouhedma formation outcrops with a very reduced thickness. Three main faults have been mapped on this seismic line and correspond to the so-called F1, F18, and F2 oriented respectively N-S, NE-SW, and E-W defined below after gravimetric study.

Section S2 characterizes the NW block of the study area and is mainly oriented N-S (**Figure 1C**). This section extends between the Northern flank of the Jebel Oum Laagol Anticline and the Southern flank of the Jebel Belkhir Anticline. From North to South,

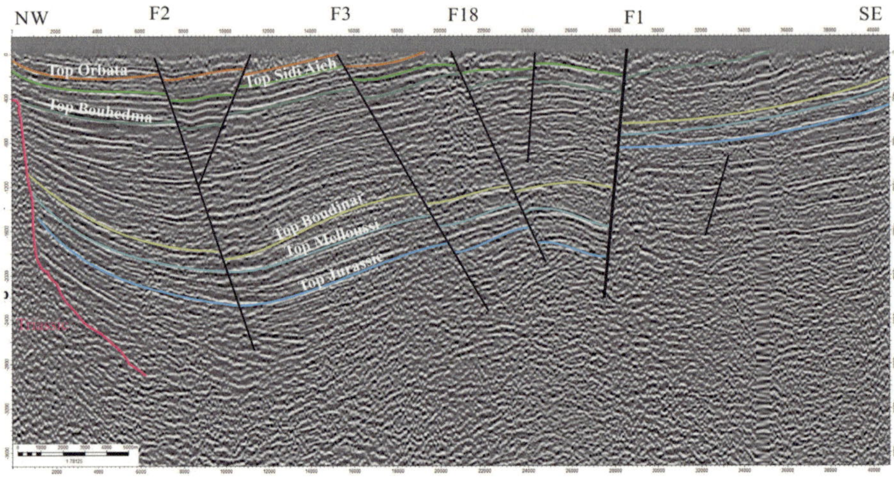

Figure 7.
NW-SE interpreted seismic line showing the deep geometry of the MHP in response to the most major fault detected by gravimetric analysis. Note the role played by the F1 fault in the differentiation of two different basins, in the SE side, and the existence of Triassic intumescence, in the NW side.

this section has well highlighted the F3 and F4 faults, which are E-W oriented and the F17 (**Figure 8**), called Hadhifa Fault, which constitutes a segment of the SAFC. Several thickness and facies variations are underlined on this N-S-oriented seismic section. Indeed, from South to North the Bouhedma formation gradually decreases in thickness, while the Sidi Aich formation records a net increase in thickness. Melloussi, Boudinar, and Orbata formations keep a constant thickness throughout the seismic line (**Figure 8**). Furthmore, two Triassic saliferous bulges characterize this seismic section denoting an intense halokinetic activity having started to individualize at least since the Barremian. This halokinetic activity coupled with the dynamics of faults F17 and F3 established a deep geometry materialized by two synclines separated by a narrow anticline.

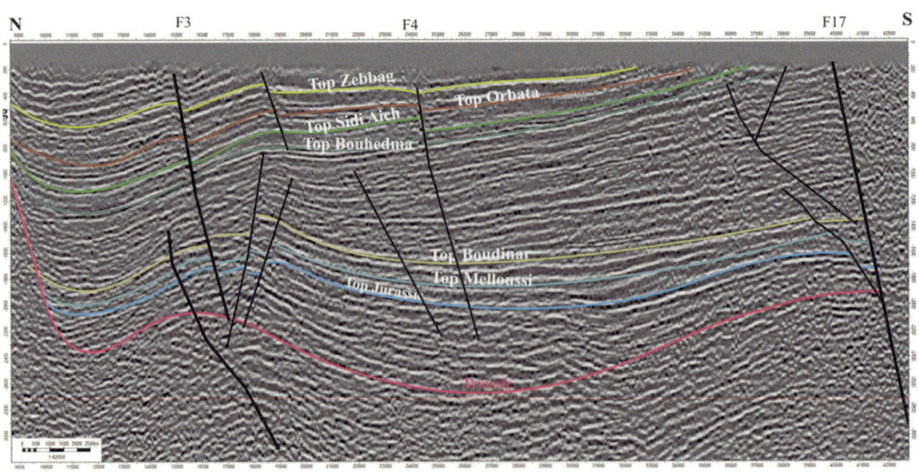

Figure 8.
N-S interpreted seismic line showing the important role played by the F17 fault in the structuration of the subsoil of MHP and the thickening of the early cretaceous series in the northern side.

Section S3 initiates from the NW flank of the Zemlet Elbeidha anticline, crosses Fejij plain, Araguib Itama, Hajri, Menzel Habib plain and continues through Sabkhet Sidi Mansour (**Figure 1C**). It shows that the subsoil of these localities is compartmentalized, from South to North, into three distinct blocks bounded by four NE–SW, E-W, and N-S oriented faults, corresponding respectively to F6, F8/F7 and F1 faults deduced from the gravimetric study (**Figure 9**). These bordering faults of hidden basins have normal slip and show small reverse bends, which could be induced by progressive structural inversions. In addition to these faults, this section shows the existence of two Triassic saliferous bulges delimiting a clear synclinal basin. The structural-stratigraphic architecture presented by this section proves that these faults, as well as the Triassic intumescences largely controlled sedimentation during the Lower Cretaceous. Indeed, South of the F7 fault, the Melloussi, Boudinar, and Bouhedma formations show a strong reduction in thickness, while the Sidi Aich formation is at its maximum thickness. Gradually toward the North, the Bouhedma formation is gaining in thickness with a small reduction above the salt structure evolving directly above the F1 fault, while a depocentre is observed between the two salt structures for this formation. This architecture denotes that, during Early Cretaceous, sedimentation is made contemporaneous to an active halokinetic rise.

Section 4 (**Figure 1C**) with a NE–SW orientation, is orthogonal of Section 1 and crosses the subsoil of Zograta, Ouali, and Aouled Khoud characterizing the NE province of the study area (**Figure 2b**). Several faults have been crossed by this section. From NE to SW, these faults correspond to the F12, F11, F13, F15, F16, and F9 previously defined by the gravimetric study (**Figure 10**). Mapping of the different horizons shows that these faults have controlled sedimentation during Early Cretaceous. The central part located between faults F15 and F16 shows that the Melloussi, Boudinar, and Bouhedma formations are very reduced in thickness, while the Sidi Aich and Orbata formations are very thick; gradually, toward the NE of fault F15 and to the SW of fault F9, the Sid Aich and Orbata formations lose thickness and the Melloussi, Boudinar, and Bouhedma formations become thicker (**Figure 10**). The deep architecture offered by this section is that of a large graben formed by the aforementioned faults testifying to a tectonic inversion of the ancient horst occurring from the Upper Barremian (**Figure 10**).

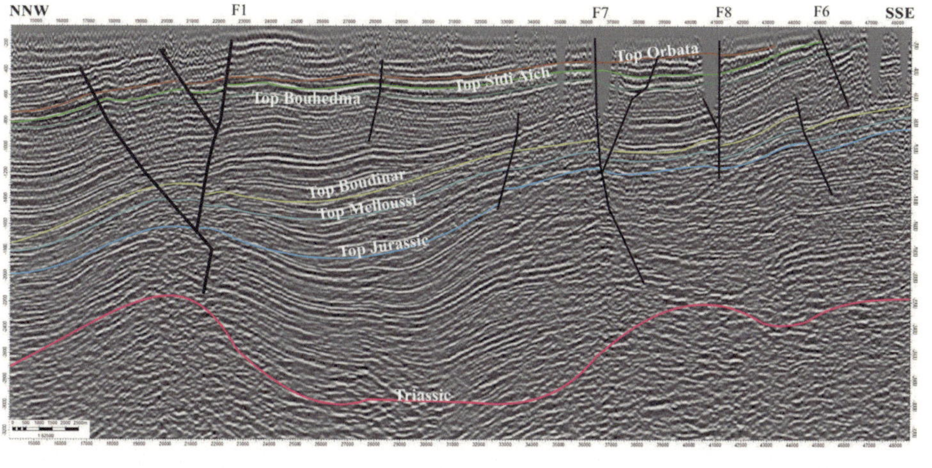

Figure 9.
NNW–SSE interpreted seismic line presenting the tectonostratigraphic role accomplished by the F1 fault and the progressive thickening of the early cretaceous formations toward the NW.

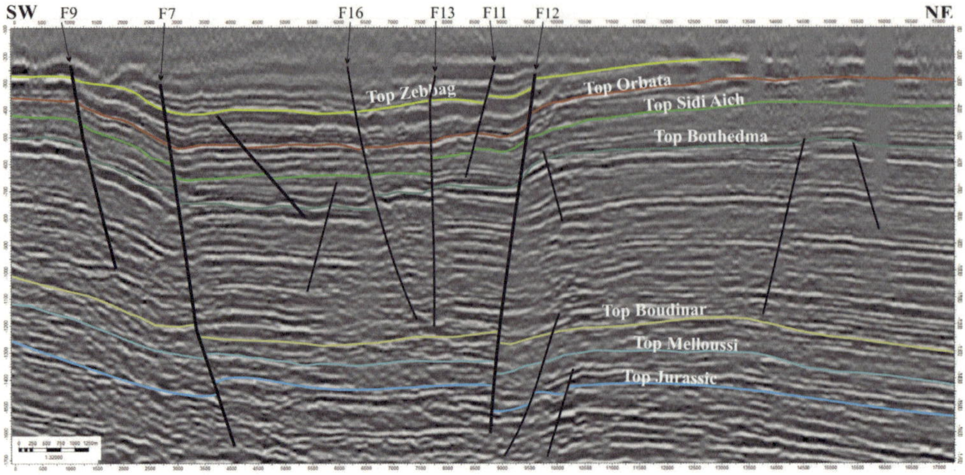

Figure 10.
NE–SW interpreted seismic line demonstrating the general deep architecture of the eastern province.

Section 5 (**Figure 1C**), with a E-W orientation, initiated from Hajri and crosses Sidi Mansour Sebkha, Mhamla, and Aoussaj plains. The different seismic horizons pointed at this section make it possible to clarify the role played by F1 fault, oriented N-S, the F17 NW-SE oriented fault representing a segment of the SAFC and ends near a buried salt structure characterizing the subsoil of the plain of Bled Segui (**Figure 11**). This seismic section shows that the deep structure crossed is compartmentalized in three blocks separated by the predicted faults. The compartment located east of the F1 fault shows that the Early Cretaceous formations are very reduced in thicknesses, but the central block exposes the thickest series. In addition, by approaching toward the salt structure, these formations of the Lower Cretaceous gradually lose thickness and show progressive bevellings. This seismic section clearly proves that the sedimentation of the Lower Cretaceous is largely controlled by the normal dynamics of the N-S and NW-SE faults and by the early halokinetic activity having, probably, started since Late Jurassic time [37, 38], on the one hand, and that the sedimentary architecture induced is that of a mega tilted block toward the West, generated by the activity of the F1 fault, on the other hand (**Figure 11**).

5.3 Facies and thickness variations

The Lower Cretaceous series are exposed only at Jebel Zemlet Elbeidha and Jebel Oum Laagol (**Figure 1C**). The Melloussi and Boudinar formations, as well as the lower part of the Bouhedma formation, are lacking and do not outcrop. They are crossed only by oil wells drilled in the region. To understand the role of the various faults, thus defined, in the distribution of these series, two transects of correlations were carried out. The first one is trended N-S (**Figure 12**), using data from oil wells, while the second is oriented NE-SW (**Figure 13**) highlighting the variations of facies and thicknesses of the outcropped series, at Jebel Zemlet Elbeidha, on both sides of the existing faults. **Figure 12** illustrates the first correlation line drawn within a N-S direction for the eastern part of the study area. From South to North, this correlation transect comprises four lithostratigraphic columns representative of the fourth petroleum drilling wells, respectively named ZB1, BHS1, BK1, and MAN1. The two

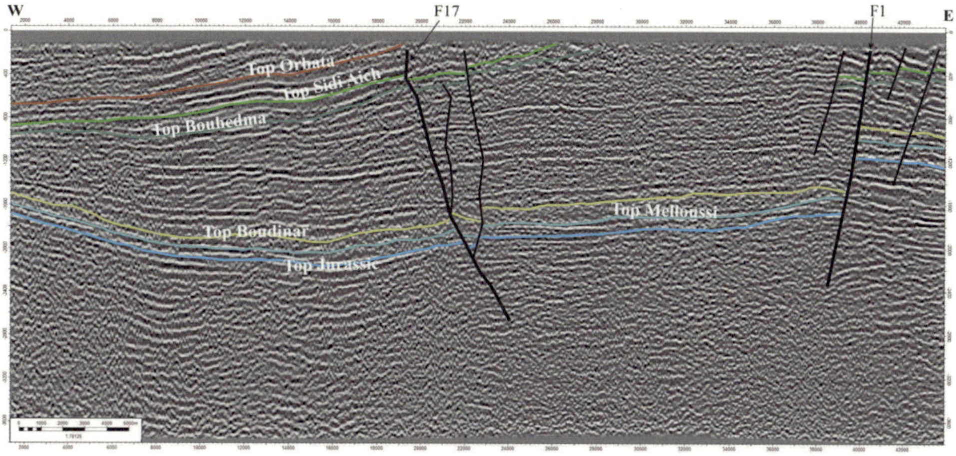

Figure 11.
E-W interpreted seismic line crossing the MHP and showing the roles of the F1 and F17 faults in the structuration of the study area.

first wells are drilled at the core of Zemlet Elbeidha anticline and at the center of the MHP. Structurally, these two wells are separated by the F6 fault characterizing the NW flank of Zemlet Elbeidha. The BK1 well is drilled at the NE termination of Jebel Belkheir. These two last wells are separated by the layout of the N-S F1 fault. MAN1 petroleum well characterizes the Southern flank of Jebel Bouhedma.

Lithostratigraphic columns studied show significant facies and thicknesses variations of the different formations attributed to the Lower Cretaceous. Indeed, from south to north, we note a gradual thickness decrease of the Melloussi, Boudinar, and lower part of the Bouhedma formation. The lithostratigraphic column of the ZB1 petroleum well intersected a Bouhedma formation three times thicker (1300 m) than

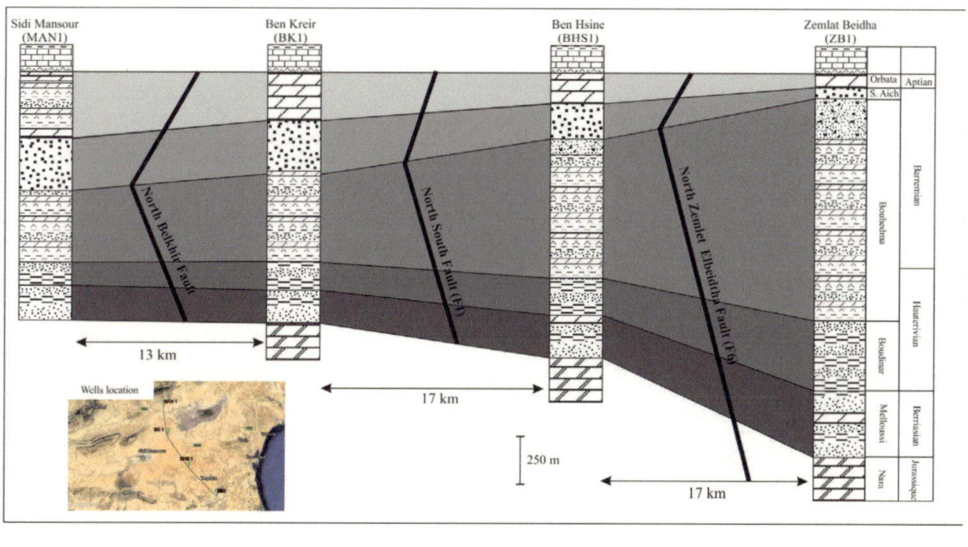

Figure 12.
NNW–SSE correlation line of the early cretaceous formations crossed by the different oil wells characterizing the study area. Noting the clear change in fault dip that occurred from late Barremian inducing a change in subsidence which becomes toward the NW.

that crossed by the MAN1 one (420 m). The sandy part, which characterizes the top of the Bouhedma formation, was encountered only at the ZB1 well with a thickness exceeding 150 m, while it is absent or very reduced in the other wells (30 m = m at BHS1). The thickness of the Melloussi and Boudinar formations make about 450 m in Zemlet Elbeidha structure, while they do not exceed to 250 m in the Belkheir and Mansour Structures.

In addition, a clear inversion of subsidence occurred during the Late Barremian-Aptian period. This inversion is materialized by a strong thickness of the Sidi Aich and Orbata formations, in the North (Jebel Belkheir and Mansour 500 m), and a progressive reduction of this thickness toward the South (Zemlet Elbidha structure), where it registers at most 100 m thick. We also note that the Orbata formation, crossed by the MAN1 petroleum well, is formed by clays and dolomites, while it is exclusively dolomitic in the other localities.

Figure 13 illustrates the second correlation line, which is mainly oriented NE-SW. It comprises six lithostratigraphic columns characterizing the Southern flank of the Zemlet Elbeidha anticline (**Figure 13A**). Each column is separated from another by a directional fault (**Figure 13A**). Significant thickness and facies variations are highlighted by this correlation line. It shows that the sandy deposits, characterizing the summit part of the Bouhedma formation, as well as the Sidi Aich sandy formation shows, at the L4 locality, the most reduced thickness. On both sides of this locality, these sandy formations recorded a progressive increase in their thickness (**Figure 13B**). This sedimentary design, during this period, is governed by the normal activities of the different defined faults inducing a horsts and grabens architecture (**Figure 13C**).

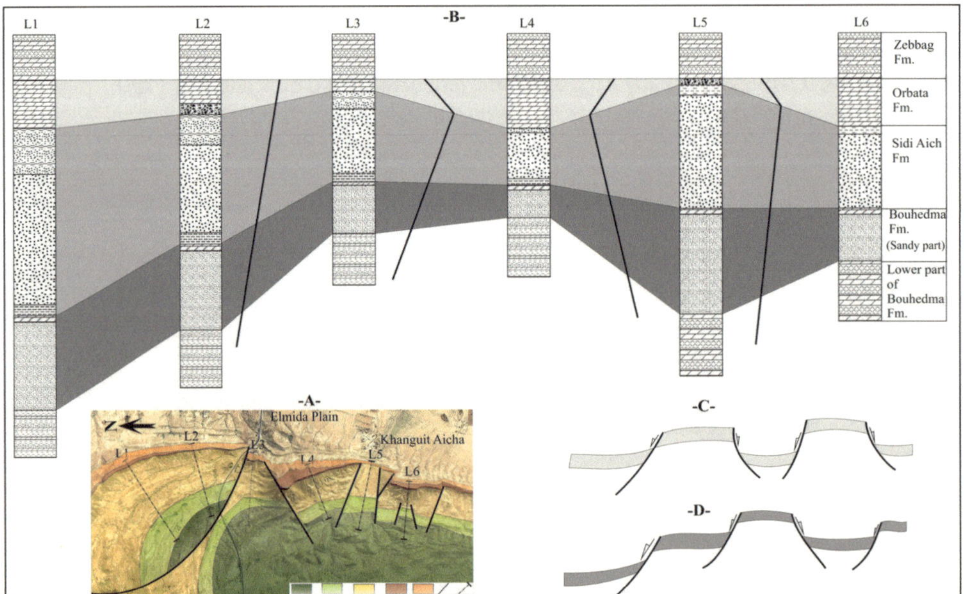

Figure 13.
Role of the different network of faults affecting the southern flank of the Zemlet Elbeidha anticline in the distribution of the Barremian and Aptian sedimentation; A- simplified geological map of the southern flank of the Zemlet Elbeidha anticline showing the location of the different lithological columns correlated; B- correlation line of the Barremian-Aptian formation characterizing this site; C and D- fault geometries during sedimentation demonstrating a clear fault dipping change from Barremian to Aptian periods.

In addition, dolomites attribute to the Orbata formation show continuous variations from one locality to another (**Figure 1B**). The lithostratigraphic column L3 exposes the most reduced series, and the column L5 demonstrates that the Orbata formation is limited to two meters of conglomerates. The thickest series has been identified in column L4. This sedimentary design is largely controlled by the normal activity of the various faults characterizing the Southern flank of Zemlet Elbeidha structure, which induce a horst and graben architecture (**Figure 13D**) but conversely with the structuration of the underlying sandy series.

5.4 Synsedimentary faulting analysis

To determine the structural evolution of the NE termination of the Southern Atlas including the MHP, during Early Cretaceous, we measured the orientations of striations on fault planes that affect different formations. Syndepositional fault populations, providing direct dating of tectonic events, have been specially analyzed. Deposits attributed to the Valanginian and the Hauterivian, defining the Melloussi and Boudinar formations, do not outcrop in our sector of study. For this reason, we will be interested to the synsedimentary faults affecting the Bouhedma, Sidi Aich, and Orbata formations, which are exposed at Jebel Zemlet Elbeidha and Jebel Oum Laagol. We present below the main sites fossilizing a tectonic evolution during the Early Cretaceous, which is materialized by synsedimentary faulting.

Red marly, and beige to yellow sand alternations, attributed to the Late Hauterivian period, outcropping at the core of Zemlet Elbeidha anticline, fossilized many syndepositional normal faults. These last are organized into two directional families (**Figure 14A**). The first family contains N40-60 trending faults, while the second one is represented by NW-SE-oriented faults. These syndepositional faults impose a horst and graben design. The numerical treatment of these fractures shows that the medium Barremian sedimentation is controlled by an extensive tectonic applying a N-S trending minimal horizontal stress (**Figure 14A'** and **B**).

Several normal synsedimentary faults (**Figure 14C** and **D**) affecting clay, dolomite, and gypsum alternations, attributed to the Hauterivian and Barremian periods and defining the Bouhedma and Sidi Aich formations, characterizing Jebel Oum Laagol show that this sedimentation was controlled by an extensive tectonics involving a N-S trending minimum stress (**Figure 14C′** and **D′**).

The red dolomitic bar, which marks the summit of the Bouhedma formation, which outcrops between Khanguit Amor and Khanguit Aicha at the Southern flank of Zemlet Elbeidha anticline, is affected by multitudes normal synsedimentary faults (**Figure 14E**). These are oriented E-W, NE-SW, and NW-SE and draw an architecture in horst and grabens. Numerical analyses of these fractures show that the Late Barremian sedimentation is controlled by an extensive tectonic applying a N-S to NNE-SSW trending minimal horizontal stress (**Figure 14E'**).

Dolomites attributed to the Aptian period and which characterize the Orbata formation are affected by normal faults trended NE-SW, NW-SE, and E-W (**Figure 14F, G**, and **H**). These faults draw a tilted blocks and a horst and graben geometries (**Figure 14H**). Numerical analyses of these faults prove that the Aptian sedimentation was subjected to an extensive tectonics applying a NE-SW minimal stress (**Figure 14**F′, G', H′, and I).

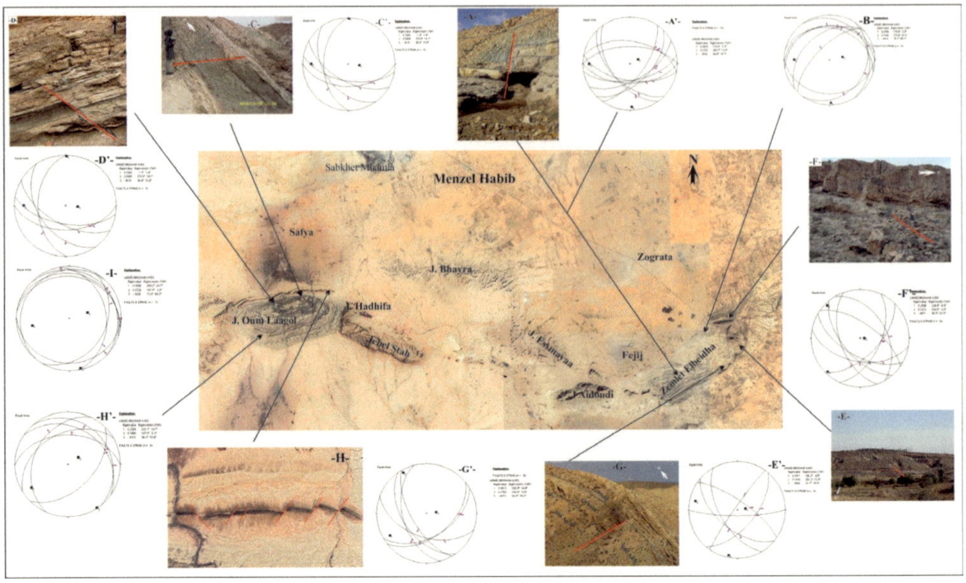

Figure 14.
Syndepositional faulting affecting the Barremian-Aptian deposits recorded at the different outcropped series in the study area (a, C, D, E, F, G and H) and characteristic stereographic projections corresponding to data measurements, Schmidt projection, lower hemisphere (B, a', C', D', E', F', G', H' and I) (the central figure shows the location of the studied sites); a, A'B) fault planes are great circles; slickenside lineations are small centrifugal arrows (normal faults), black squares with index 1: σ_3, black squares with index 2: σ_2 black squares with index 3: σ_1.

6. Discussion: plaleostructuration and hydrogeological implication

6.1 Basin shapes

The compilation of morphological, gravimetric, and seismic data shows that the MPH subsoil is structured in diamond-shaped to rectangular blocks shaped by the combination of N-S and NW-SE oriented major faults with E-W and NE-SW second-order faults. The NW-SE and E-W faults, which dominate the subsoil of the western province architect this zone into two subsiding basins of lozenge-shaped forms separated by a raised band (**Figure 15A**). This arrangement is that of a horst and graben system, during periods of sedimentation, evolving after tectonic inversion into two large synclines separated by a narrow anticline (**Figure 15B**). To the east and near the fault F1, these two synclines are wide and deep. They gradually narrow toward the West near Jebel Berda and are very disturbed by the halokinetic activity molded on the F17 master fault. Geometrically, we point out the progressive deepening of the Lower Cretaceous series from West to East and from South to North. This architecture could affect the underground flow of water contained in the porous and permeable series of the Lower Cretaceous. The overall collapse of the western province is due to the F1 Fault in comparison with the eastern province (**Figure 15**). This last is constantly controlled by N-S and NE-SW trending faults. Thin formations characterizing this province compared to that identified at the western province drawed a large synclinal basin (**Figure 15A**) of sub-rectangular shape delimited by two major N-S oriented faults, in the East and West parts, and by two NE-SW trending fault, in the North and South borders. This buried structure is continuously affected by second-order faults. Geometrically, this highly fractured syncline structure shows a regular slope toward the SE.

DOI: http://dx.doi.org/10.5772/intechopen.112209

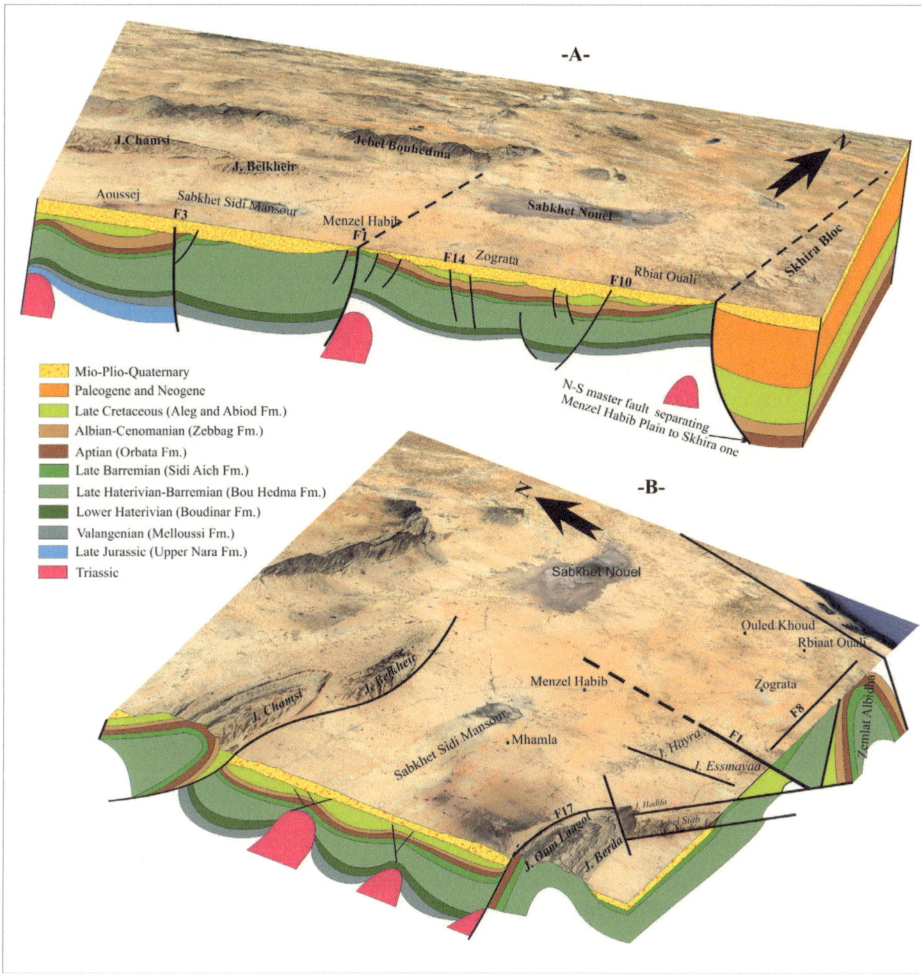

Figure 15.

General basin shapes of the subsoil of MHP; A- E-W cross section; E-W geological section showing the deep architecture of the study area and the roles of the N-S and NE–SW faults in the individualization of this architecture; B- N-S section showing the deep structure of the study area in relation to the NW-SE and E-W fault systems.

In addition to these different fault networks, which determine the deep geometry of the Menzel Habib plain, there is an intense halokinetic activity, which creates bulges localized essentially in the layout of large faults (**Figure 15**). The most important salt structures are located below the fault F1, F17, and F8, which characterize the subsoil of Menzel Habib Village-Jebel Belkheir, the North flank of Jebel Oum Laagol and the subsoil of Sabkhet Sidi Mansour-Mhamla, respectively.

6.2 Paleo structuring

The multiscale approach adopted to understand the geodynamic evolution of the MPH, based on the morphological analysis, the interpretation of gravity and seismic data coupled with the inventory of facies and thickness variations of the Lower Cretaceous formations and the examination of the synsedimentary faults prove that this plain is dominated, during the Early Cretaceous time, by a N-S to NE-SW extensive dynamics. This tectonic event results in the individualization of several basins of variable size and extent, evolving at the expense of a complex network of faults. This architecture has greatly influenced the hydrogeological evolution of this plain.

Morphological analysis allowed to note that the MHP is differentiated from East to West into two different hydrological provinces separated by a N-S oriented bulge, which is traced in the center of this plain and serves as a water sharing (**Figure 3**). It corresponds to a surface response of the F1 fault (**Figures 6** and **9**) elongated in the same direction. Regionally, this morphologic protuberance constitutes the extension toward the South of the morphostructural line, elongated from Hammem Lif, in the North, to Mezzouna, in the South, which is molded on a N-S paleogeographic fault, commonly called N-S Axis. Although, morphologically not very prominent, this central sub-meridian uplift reflects the effects of a jerky geodynamic evolution and plays a very important role in the hydrogeological differentiation of the Menzel Habib basin.

The most impressive gravimetric differentiation is that managed by an N-S trending fault separating an eastern basin bearing the signature of dense series and a western basin characterized by a significant mass deficit (**Figure 5A**). This fault coincides with that was identified by morphological analysis and corresponds to the southernmost branch of the N-S axis. The residual map, the vertical and horizontal gradient maps, as well as the tilt angle map show that the subsoil of MHP is shaped into several E-W and NE-SW trending basins, which are largely disturbed by N-S, NE-SW, E-W, and NW-SE oriented faults (**Figure 6B**).

The interpretation of the available seismic lines confirmed the existence of all the faults detected by the morphological and gravimetric analyzes. These faults had a major impact on the distribution of the various geological formations that characterize the study area during early extensive periods and guided its structural differentiation during later orogenic phases. The F1 and F17 faults, which present a N-S and NW-SE direction, respectively, are the most important accidents that have contributed to the shaping of this study area. In addition, the geodynamic evolution of the study region is also due to the E-W and NE-SW faults, currently exhibiting reverse and sinistral reverse activities respectively (**Figures 6–8**).

During the Early Cretaceous period, the extensive dynamic governing the Tunisian margin and applying a N-S to NE-SW minimal stress [39] activates the F1 fault with a normal dynamic allowing the uplift of the eastern block and the subsidence of the western one (**Figures 16** and **17A**). This architecture promotes the accumulation of a thick and continuous series to the West and a reduced one in the eastern province (**Figure 4A**).

From North to South, Burollet [40], Abbès et al. [41], Boccaletti et al. [42], Dolglioni et al. [43], Martinez et al., [44], and [45] demonstrate that this architecture has been proven on all the different branches of the N-S Axis. Furthermore, this master lineament constitutes a sedimentary limit between two domains, which have different palaeogeographic settings, the Atlas zone, with very thick Lower Cretaceous formations, and the Eastern platform, characterizing a slow subsidence, during this period. Arfaoui et al., [45] denote that this alignment formed a palaeogeographic feature since the Jurassic as the reductions or condensations of the sedimentary sequences often show gaps and discordances.

In addition, and in response to this extensive tectonic, the F17 fault, representing the t branch of the SAFC, is activated as normal dynamic with a NE dipping (**Figure 17A**). It results accumulation of continuous and thick series in the NE part (3000 m thick of deposits) and reduced ones in the SW province in which the total thickness of the Jurassic

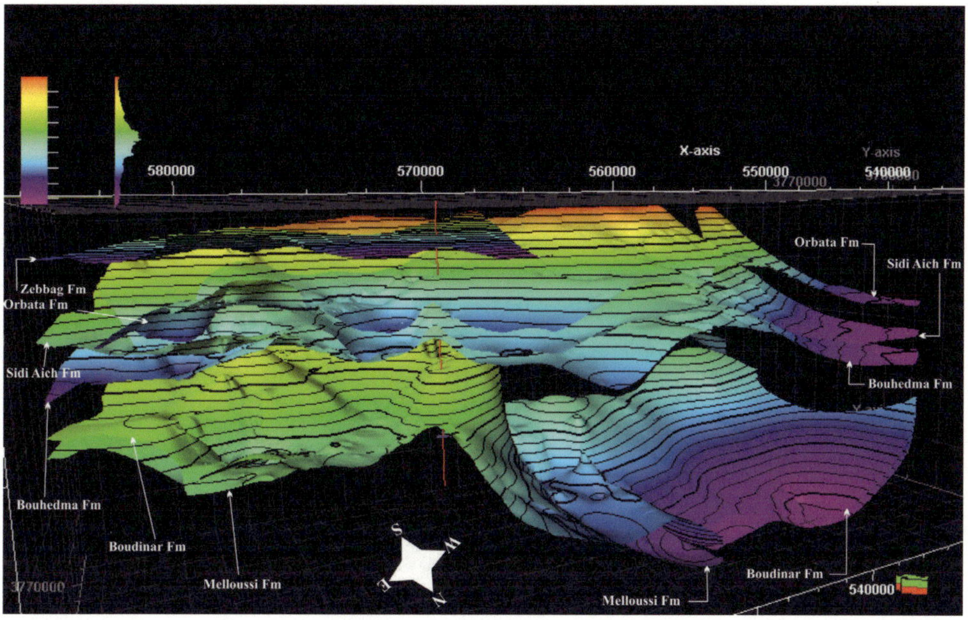

Figure 16.
Superposition of the isochronous maps of the different formations characterizing the subsoil of MHP showing the role played by the F1 fault in the paleostructuration of the study area. Noting essentially the big variation in thickness of Melloussi, Boudinar and Bouhedma formations on both sides of this fault.

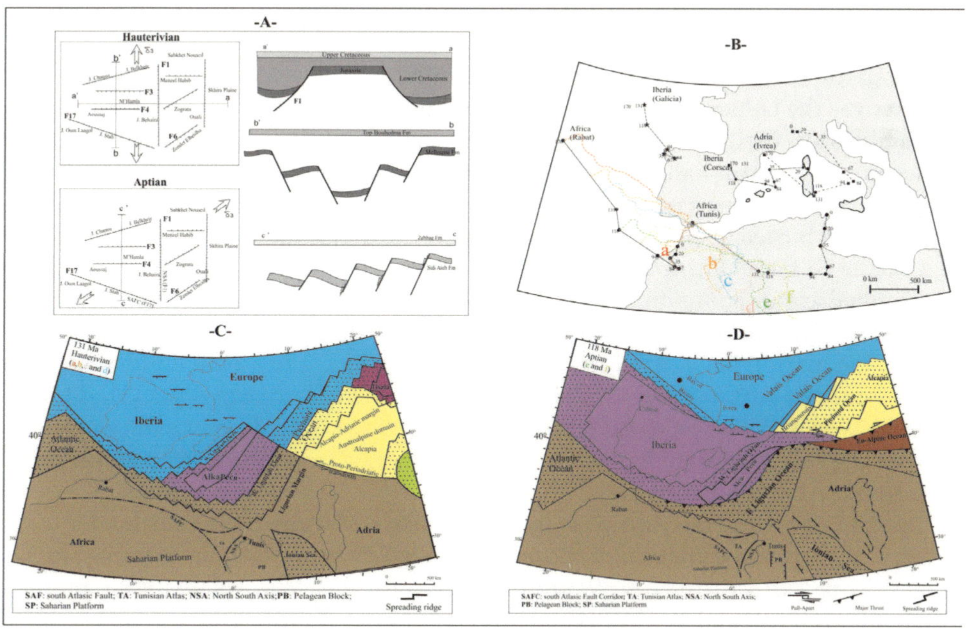

Figure 17.
Evolution of the Menzel Habib plain in response to the N-S and NE–SW extensive tectonics during early cretaceous. A- two evolutionary models of Menzel Habib plain in relation to the most important fault which affect the study area; B- relative motion of Africa with respect to Europe. Six steps (a, b, c, d, e and f with different colors) of displacement of Tunis summarize the relative trajectory of Africa during early cretaceous [11]; C and D plate tectonic maps of alpine Tethys and the western embayment of Neotethys during Hauterivian and Aptian periods respectively.

and Early Cretaceous series does not exceed 1600 m [46]. This architecture has been highlighted and proven by Said et al., [47, 48] and Arfaoui et al., [45]. These two authors have confirmed that these major faults signed, during the Late Triassic to Early Cretaceous, the initiation and the accommodation of the Tethyan rifting inducing a complex system of horst and graben structures.

The structural associations characterizing these two major palaeogeographic faults (N-S axis and Gafsa-Hadhifa accident) lead to consider them as first-order structures to which the MHP owes its existence and its evolution since the first stages of Tethyan rifting.

Second, the stratigraphic evolution of the MHP depends on the NE-SW fault, which runs along the NW flank of Zemlet Elbeidha, and the E-W fault, which borders Jebel Belkheir and Jebel Chamsi on the south side. During the Valanginian-Hauterivian, these two faults dip toward the South, while they dip toward the North during the Barremian to Aptian period (**Figure 17A**).

In response to the Early Cretaceous NE-SW to N-S extensive tectonics, these faults compartmentalize the subsoil of Sidi Mansour-Mhamla-Aoussaj into several rhombic basins of different geometry and size with differential migration of the subsidence. This architecture is that of a large graben made by the Hadhifa Fault (F17), which dips toward the North and the major E-W fault bordering the Jebel Chamsi and Jebel Belkheir anticline's, and which dips to the South (**Figure 17A**). On the other hand, the F1 fault and the major fault separating the MHP from the plain of Skhira, which plunge toward the West and toward the East, respectively delimit a great horst characterizing the subsoil of Zograta-Rbiat Ouali (**Figure 17A**).

Detailed analysis of the different syndepositional faults fossilized in the Early Cretaceous series proves confirm that this extensive tectonic applies two successive minimum stress oscillating between N-S and NE-SW poles. The first direction of stress is attributed to the Valanginian to Barremian period, while the second one characterizes the Upper Barremian to Aptian time.

In addition to their synsedimentary activity these faults guided, during the Lower Cretaceous, significant halokinetic activity materialized by Triassic bulges. Some of these Triassic intumescences reach the surface, case of Jebel Hadhifa (**Figure 1**), others remain buried and mark out the layout of the major faults, case of the Triassic bulge on the F1 fault (**Figure 9**). These salt bodies implanted on the master faults give an additional argument on the role played by these faults in the geodynamic evolution of that MHP. Chronologically, the N-S extensional tectonics recorded at this region of the Tunisian margin has started since the Triassic period. It has been highlighted at the Southeastern part of Tunisia [5, 35, 49] at the Central Tunisian Atlas [31, 44, 50] and at the Northern Tunisian Atlas [6, 28, 29]. This extensional event evolved in relation with the sub-meridian Tethyan extensional framework, which affected the whole of the North-African margin [51].

This extensive dynamic, which ruled the Tunisian Margin during the Lower Cretaceous, has been widely studied and proven in Libya [52–54], in Italy [55], in Algeria [56–58], in Morocco [8, 59] and in Spain [60]. It is responsible for an intense rifting applying N-S and N-–SW faults favoring the development of several NE-SW trending grabens filled with continental deposits. Moreover, this distension is accompanied by a very important basaltic magmatism [61].

The extensive tectonics, which govern the Tunisian margin as well as the neighboring margins and which applies two minimal stress, trended NS and NE-SW, representative of the Valanginian-Hauterivian-Lower Barremian and the Upper Barremian-Aptian times respectively, translate the echoes of a jerky evolutionary motion of Africa and Iberia in respect to Eurasia plates during Tethyan and Atlantic spreading (**Figure 17B**) [11, 62, 63]. In fact, the first geodynamic stage of extension is contemporaneous to the northward propagation of the Atlantic Ocean spreading along the western margin of Iberia and the sinistral displacement of Africa relative to Europe (**Figure 17B** and **C**) [64, 65], which produce the abandonment of the Ligurian–Tethys corridor and the opening of the Bay of Biscay along the already rifted Pyrenean basin with a concurrent counterclockwise rotation of Iberia [11, 66, 67]. The second geodynamic stage characterized by NE-SW dynamics results from the Northward propagation of the Central to North Atlantic, which led to the individuation and eastward motion of the Iberian microplate and Africa with respect to Europe ([11]; Berra and Angiolini [68] (**Figure 17B** and **D**) Menant et al. [69], and Ye et al. [70] Guerrera et al. [71]. After this significant plate reorganization and increase of the Iberia-Africa-Eurasia convergence velocity at ~118 Ma [54, 72], results the onset of subduction of the eastern Ligurian part of Alpine Tethys along the Iberian–African plate boundary, which continuous with intracontinental subduction and Eo-alpine orogenesis (**Figure 17D**) at the northern tip of Adria, where Alcapia and Adria converged obliquely during the same time period [11, 54, 70].

For the MHP case, the change of stress field from a N-S pole to a NE-SW one represents the transition between the opening of the South Atlantic, the sinistral drift toward the SE of Africa in relation to Eurasia and the beginning of subduction of the Neo-Tethys [63, 73–75], and the beginning of the opening of the Eastern Mediterranean accompanied by the NW migration of Apulia and the development of Back-arc basins above the northern subduction zone, forming oceanic crust now flooring the Black Sea and the Caspian Sea [54, 76–80].

6.3 Hydrogeological implication

The geometry of the various buried structures characterizing MHP, coupled with the high porosity and permeability of the different sandy and dolomitic series, attributed to the Early Cretaceous time, prove that the subsoil of MHP exhibits a very developed deep aquifer system characterizing the eastern end of the Southern Tunisian Atlas. The inventory of the various water wells, carried out in 2017, confirms that the MHP subsoil contains a very important multilayer aquifer system. These layers correspond to the sandy and dolomitic deposits of the Melloussi formation (300 m), the sandy series of the Boudinar formation (250 m), the top sandy layers of the Bouhedma formation (150 m), the sandy deposits of the Sidi Aich formation (100 m), and the dolomitic dale of Orbata formation (50 m) (**Figure 2**). The recharge of this aquifer is done either by direct infiltration from the outcropped formations at Jebel Zemlet Elbeidha and Jebel Oum Laagol, which dips toward the NW or by vertical infiltration through the various faults affecting the surface from the numerous sabkhats characterizing this zones.

Therefore, the available piezometric maps ([25, 81]; CRDA Gabès), the structural model (**Figures 15** and **16**), the hydrographic network (**Figure 3**), and the underground water flow direction enabled us to draw a detailed piezometric map of the deep aquifer characterizing the MHP. This map shows that this area is compartmentalized into two hydrogeological compartments separated by a water-sharing line corresponding to the layout of the F1 master fault parallel to the dividing line of the surface water. The piezometric map of the eastern compartment clearly shows that the underflow is directed from west to east taking the same direction as the flow of Oued Rimth. In the western compartment, the underground flow is multidirectional

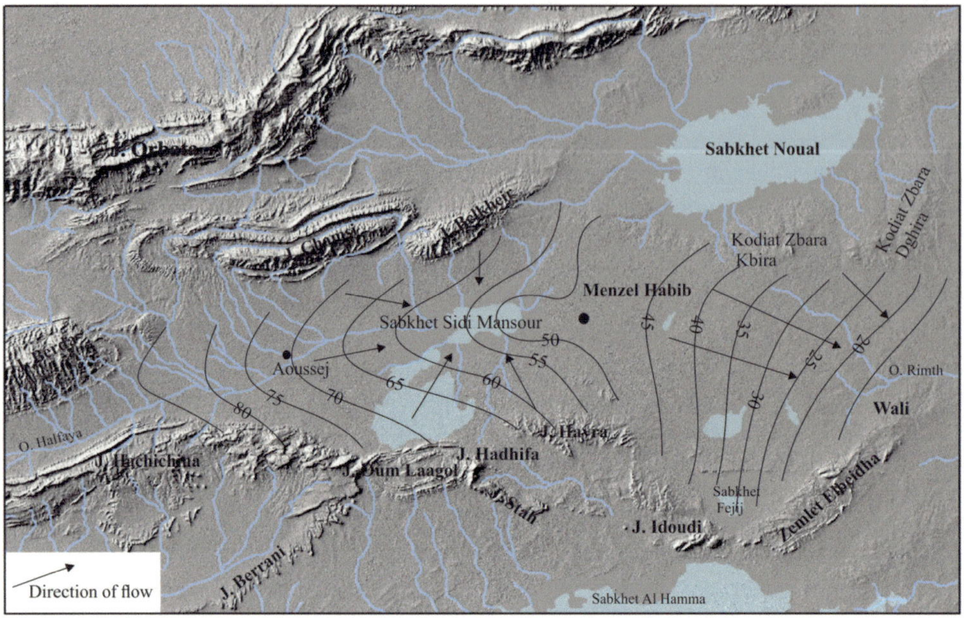

Figure 18.
Piezometric map of deep groundwater table ([25, 81]; CRDA Gabès, 2019).

converging toward the basement of Sabkhet Sidi Mansour imitating the flow of the different wadis flowing into this Sebkha (**Figure 18**).

Consequently, the deep aquifer system of MHP appears by its structural arrangement between the different structures of the central and Southern Tunisian Atlas, the Eastern platform, and the Jeffara by its geometry, as well as by its complex geodynamic evolution in relation to the major faults (NSA and SAFC), as an independent hydrogeological entity and different from other aquifers characteristic of neighboring areas. This multilayered aquifer of MHP is separated from the North Gabès Aquifer (NGA) by the N-S (F1) and the Al Mida NE-SW faults, from the Continental Intercalaire Aquifer (CI) of El Hamma-Kebili by the Hadhifa NE-SW Fault (F17), from the Guettar-Gafsa Nord Aquifer by the Chamsi-Belkheir anticlines and by the Gafsa Fault (Branch of the SAFC), from Sfax Aquifer by a NW-SE fault, which borders Sabkhet Noual and from the coastal Aquifer of Gabès by the N-S fault parallel to the F1 fault (**Figure 18**).

7. Conclusions

The main results that can be enhanced from our research on structural-stratigraphic field investigations and the analysis of gravimetric and seismic data, targeting the Menzel Habib plain, which represents a hinge zone between the major structural features of the Southern Tunisian margin, can be summarized as follows:

- The hydrographic and morphologic approach proves that the MHP is subdivided into two distinguished hydrologic basins separated by a N-S topographic bulge, which expresses the surface response of the deep activity of the F1 fault.

- The synsedimentary faults affecting the various formations of the Early Cretaceous, as well as the variations in thickness and facies of these formations, on both sides of the major faults, prove that this period was controlled by an extensive tectonics applying a minimal stress oscillate between a N-S pole and a NE-SW pole. The Valanginian-Hauterivian-Lower Barremian period, with a N-S extensive event, constitutes the continuation of the long rifting epoch, which began since the Late Triassic time and which is responsible for the installation of several basins evolving in respect to N-S and NW-SE normal faults. From the Late Barremian period the minimum stress becomes NE-SW. It is responsible for the amplification of the subsidence of the basins following a horst and graben or tilted block architecture. The N-S, E-W and NW-SE faults are the most important and the most imposing syndepositional faults to ensure this evolution.

- The deep architecture of MHP, deduced from the available gravity maps, is that of two basins with different densities, separated by a major N-S oriented fault (F1). The eastern compartment is formed by a thin and dense series, while the western basin is characterized by a significant mass deficit. The residual map, the vertical and horizontal gradient maps as well as the tilt angle map show that the subsoil of MHP is shaped into several E-W and NE-SW trending basins, which are largely disturbed by N-S, NE-SW, E-W, and NW-SE oriented faults.

- The available seismic lines clearly confirmed the existence of all the faults detected by the gravimetric and morphologic analysis. The most important faults highlighted by the seismic interpreted lines are the F17 fault, representing the most eastern branch of the SACF, and the F1 fault, which constitutes the southern segment of the NSA.

- The deep architecture of the MHP appears as a series of basins evolving to the detriment of a dense network of faults and showing varying thicknesses of the Lower Cretaceous formations. These basins constitute, by their geometries and by the composition of the various formations which form them, very developed deep aquifers which will be able to store appreciable quantities of water. These aquifers are totally independent from the usual ones known in the surrounding areas of the Southern Tunisian margin. The N-S and the NW-SE trending faults, characterizing the F1 and F17 faults, and the major various anticlines (Jebel Zemlet Elbeidha, Jebel Aidoudi, Jebel Oum Laagol, Jebel Chamsi and Jebel Belkheir) contribute to the specificities of these deep aquifers.

Author details

Mohamed Ben Chelbi
Faculty of Sciences of Tunis, Laboratory of Geodynamic, Geonumeric and Georessources (L3G) LR18ES37, University of Tunis El Manar, Tunisia

*Address all correspondence to: med.benchelbi@gmail.com

References

[1] Bodin S, Petitpierre L, Wood J, Elkanouni I, Redfern J. Timing of early to midcretaceous tectonic phases along North Africa: New insights from the Jeffara escarpment (Libya–Tunisia). Journal of African Earth Sciences. 2010;**58**(3):489-506

[2] Doglioni C, Mongelli F, Pieri P. The Puglia uplift (SE Italy): An anomaly in the foreland of the Apenninic subduction due to buckling of a thick continental lithosphere. Tectonics. 1994;**13**(5):1309-1321. DOI: 10.1029/94TC01501

[3] Guiraud R, Bosworth W, Thierry J, Delplanque A. Phanerozoic geological evolution of the northern and Central Africa. Journal of African Earth Sciences. 2005;**43**:83-143

[4] Piqué A, Tricart P, Guiraud R, Laville E, Bouaziz S, Ait AM, et al. The Mesozoic-Cenozoic atlas belt (North Africa). An overview. Geodynamica Acta. 2002;**15**(3):185

[5] Raulin C, Frizon de Lamotte D, Bouaziz S, Khomsi S, Mouchot N, Ruiz G, et al. Late Triassic–early Jurassic block tilting along E–W faults, in southern Tunisia: New interpretation of the Tebaga of Medenine. Journal of African Earth Sciences. 2011;**61**(2011):94-104

[6] Ben Chelbi M, Melki F, Zargouni F. Précision sur l'évolution structurale de l'Atlas septentrional de Tunisie depuis le Crétacé (Bassin de Bir M'Cherga). Echos d'une évolution polyphasée de la marge tunisienne dans son cadre méditerranéen. Africa Geosciences Review. 2008;**15**(3):229-246

[7] Ben, Chelbi M. Geodynamic evolution of the Tunisian margin during the Albian–Cenomanian: Structural evidence of the Austrian orogenic phase and the early tectonic inversion of the Tunisian atlas. Journal of the Geological Society. 2021;**178**(3). DOI: 10.1144/jgs2019-195

[8] Piqué A, Ait Brahim L, Ait Ouali R, Amrhar M, Charroud F, Gourmelen C, et al. Evolution structurale des domaines atlasique du Maghreb au Méso-Cénozoïque; le rôle des structures héritées dans la déformation du domaine atlasique de l'Afrique de Nord. Bulletin de la Société Géologique France. 1998;**169**(6):797-810

[9] Zouari H, Turki MM, Delteil J, Stephan JF. Tectonique transtensive de la paléomarge tunisienne au cours de l'Aptien-Campanien. Bulletin de la Société Géologique France. 1999;**170**(3):295-301

[10] Ge M, Guiting H. Mechanism of the Mesozoic African rift system: Paleostress field modeling. Journal of Geodynamics. 2019;**132**:101655. DOI: 10.1016/j.jog.2019.101655

[11] Handy MR, Schmid SM, Bousquet R, Kissling E, Bernoulli D. Reconciling plate-tectonic reconstructions of alpine Tethys with the geological–geophysical record of spreading and subduction in the Alps. Earth-Science Reviews. 2010;**102**(3-4):121-158. DOI: 10.1016/j.earscirev.2 010.06.002

[12] Ghedoui R, Deffontaines B, Rabia MC. Neotectonics of coastal Jeffara (southern Tunisia): State of the art. Tectonophysics. 2016;**676**(26):211-228. DOI: 10.1016/j.tecto.2015.11.032

[13] Rosenbaum G. Geodynamics of oroclinal bending: Insights from the Mediterranean. Journal of Geodynamics. 2014;**82**:5-15. DOI: 10.1016/j.jog.2014.05.002

[14] Ben Chelbi M, Melki F, Zargouni F. Mode de mise en place des corps salifères de l'Atlas septentrional. Exemple de l'appareil de Bir Afou. Comptes Rendus Géosciences Paris. 2006;**338**:349-358

[15] Ben Chelbi M, Kamel S, Harrab S, Rebaï N, Melki F, Meghraoui M, et al. Tectonosedimentary evidence in the Tunisian atlas, Bou Arada trough: Insights for the geodynamic evolution and Africa–Eurasia plate convergence. Journal of the Geological Society. 2013;**170**:435-449. DOI: 10.1144/jgs2012-095

[16] Ghanmi M, Bahrouni A, Ghanmi M, Zargouni F. Aptian-Albian boundary in central southern atlas of Tunisia: New tectono-sedimentary facts. Journal of African Earth Sciences. 2017;**132**:27-36. DOI: 10.1016/j.jafrearsci.2017.04.030

[17] Gharbi M, Masrouhi A, Espurt N, Bellier O, Amari EA, Ben Youssef M, et al. New tectono-sedimentary evidences for Aptian to Santonian extension of the cretaceous rifting in the northern Chotts range (southern Tunisia). Journal of African Earth Sciences. 2013;**79**:58-73. DOI: 10.1016/j.jafrearsci.2012.09.017

[18] Khomsi S, Bédir M, Soussi M, Ben Jemia MG, Ben Ismail-Lattrache K. Mise en évidence en subsurface d'événements compressifs Éocène moyen–supérieur en Tunisie orientale (Sahel) : généralité de la phase atlasique en Afrique du Nord. C. R. Geoscience. 2006;**338**(2006):41-49

[19] Masrouhi A, Ghanmi M, Ben Slama MM, Ben Youssef M, Vila JM, Zargouni F. New tectono-sedimentary evidence constraining the timing of the positive tectonic inversion and the Eocene Atlasic phase in northern Tunisia: Implication for the north African paleo-margin evolution. Comptes Rendus Geoscience. 2008;**340**:771e778

[20] Zouaghi T, Bedir M, Inoubli MH. 2D seismic interpretation of strike-slip faulting, salt tectonics, and cretaceous unconformities, Atlas Mountains, Central Tunisia. Journal of African Earth Sciences. 2005;**43**:464-486

[21] Observatoire Sahara et Sahel (OSS). Système aquifère du Sahara Septentrional: une conscience de bassin. Volume 2, hydrogéologie. Observatoire Sahara et Sahel. 2003. http://www.csf-desertification.org/index.php/bibliotheque/doc_details/39-bied-charreton-marc--2003-le-systeme-aquifere-du-saharaseptentrional-qune-conscience-de-bassin)

[22] Mejri F, Burollet PF, Ben Ferjani A. Petroleum geology of Tunisia, Mémoir de l'ETA. 2006;n°22:233. Tunisia

[23] M'Rabet A, Mejri F, Burollet PF, Memmi L, Chandoul H. Catalog of type sections in Tunisia. Cetaceous Members in ETAP. 1995;n°8A:123

[24] Ben Alaya M, Zemni T, Mamou A, Zargouni F. Acquisition de salinité et qualité des eaux d'une nappe profonde en Tunisie: approche statistique et géochimique. Hydrological Sciences Journal. 2014;**59**(2):395-419. DOI: 10.1080/02626667.2013.870663

[25] Bencheikh N, Zouari K, Abidi B. Application of chemical and isotopic methods to study of the hydrodynamic relationship between Sfax and Djeffara inthe northern Gabes aquifers (southeastern Tunisia). Hydrological Sciences Journal. 2012;**57**(8):1662-1671. DOI: 10.1080/02626667.2012.717699

[26] Kamel S, Dassi L, Zouari K, Abidi B. Geochemical and isotopic investigation of the aquifer system in the Djerid–Nefzaoua basin, southern Tunisia. Environmental Geology. 2005;**49**:159-170

[27] Kamel S, Ben Chelbi M, Jedoui Y. Investigation of sulphate origins in the Jeffara aquifer, southeastern Tunisia: A geochemical approach. Journal of Earth System Science. 2013;**122**(1):15-28

[28] Boughdiri M, Sallouhi H, Maâlaoui K, Soussi M, Cordey F. Calpionellid zonation of the Jurassic–cretaceous transition in north-Atlasic Tunisia. Updated upper Jurassic stratigraphy of the 'Tunisian trough' and regional correlations. Comptes Rendus Geoscience Paris. 2006;**338**:1250-1259

[29] Morgan MA, Grocott J, Moody RTJ. The structural evolution of the Zaghouan-Ressas structural belt, northern Tunisia. Geological Society, London, Special Publication. 1998;**132**:405-422

[30] Sekatni N, Fauré P, Alouani R, Zargoun F. Le passage Lias–Dogger de la Dorsale de Tunisie septentrionale Nouveaux apports biostratigraphiques. Âge Toarcien supérieur de la distension téthysienne. Comptes Rendus Palevol. 2008;7:185-194

[31] Soussi M. New Jurassic lithostratigraphic chart for the Tunisian atlas. Geobios. 2003;**36**:761-773

[32] Melki F, Zouaghi T, Ben Chelbi M, Bedir M, Zargouni F. Tectono-sedimentary events and geodynamic evolution of the Mesozoic and Cenozoic basins of the alpine margin, gulf of Tunis, North-Eastern Tunisia offshore. Comptes Rendus Geoscience. 2010;**342**:741-753. DOI: 10.1016/j.crte.2010.04.005

[33] Vila JM, Ben Youssef M, Charrière A, Chikhaoui M, Ghanmi M, Kammoun F, et al. Découverte en Tunisie au SW du Kef de matériel triasique interstratifié dans l'Albien: extension du domaine à «glacier de sel» sous-marin des confins algéro-tunisiens. Comptes Rendus Académie de Sciences Paris. 1994;**318 série II**:109-116

[34] Tlig S, Er-Raoui L, Ben Aissa L, Alouani R, Tagorti MA. Tectogenèses alpine et atlasique: deux évènements distincts dans l'histoire géologique de la Tunisie. Corrélation avec les évènements clés de la Méditerranée. Comptes Rendus Académie de Sciences Paris, t.1991;312(Série II):pp. 295-301

[35] Bouaziz S, Barrier E, Soussi M, Turki MM, Zouari H. Tectonic evolution of the northern African margin in Tunisia from paleostress data and sedimentary record. Tectonophysics. 2002;**337**:227-253

[36] Midassi MS. Regional Gravity of Tunisia. Masters Thesis,. USA: Univ. South Carolina; 1982. p. 125

[37] Tanfous Amri D, Bédir M, Soussi M, Azaiez H, Zitouni L, Inoubli MH, et al. Halocinèse précoce associée au rifting jurassique dans l'Atlas central de Tunisie (région de Majoura-El Hfay). Comptes Rendus Geoscience. 2005;**337**(7):703-711

[38] Zouaghi T, Inoubli MH, Bédir M. Contribution of seismic velocities to the structural and the lithostratigraphic studies: Salt-intruded corridor ceiling and lower Turonian Beida anhydrite deposits outline in central-southern atlas of Tunisia. Comptes Rendus Geoscience. 2007;**339**:13-23

[39] Ben Chelbi Mohamed. Early Cretaceous Tectonostratigraphic Evolution of the Southern Tunisian Margin Based on Gravity, Seismic and Potential Field Data: New Insights into a Geodynamic Evolution in a Tethyan and Mesogean Rifting Context. Journal of Earth Science. 2023;**34**(3):879-899. DOI: 10.1007/s12583-021-1540-x. Available from: http://en.earth-science.net

[40] Burollet PF. Contribution à l'étude stratigraphique de la Tunisie centrale. Annales Mines et Géololgie Tunis. 1956;**18**:350

[41] Abbes C, Turki MM and Truillet R. Un élément structural nouveau dans l'Atlas tunisien: le contact tangentiel décakilometrique à vergence Ouest des Djebels Ousselat et Bou Dabous (Axe Nord-Sud - Tunisie). Comptes rendus de l'Académie des Sciences, t. 1981; 292, série II: 473-476

[42] Boccaletti M, Cello G, Tortorici L. Structure and tectonic significance of the north south Axis of Tunsia. Annales Tectonicae. 1988;II(n° 1):12-22

[43] Doglioni C, Bosellini A, Frare MC, Dhaha F, Ben Said E. Annali Dell Universita di Ferrara, Scienze della Terra. 1990;2(n°5):77-94

[44] Martinez C, Chikhaoui M, Truillet R, Ouali J, Creuzot G. Le contexte géodynamique de la distension albo-aptienne en Tunisie septentrionale et centrale: structuration éocrétacée de l'Atlas tunisien. Eclogae Geologicae Helvetiae. 1991;**84**:61-82

[45] Arfaoui M, Reid A, Inoubli MH. Evidence for a new regional NW–SE fault and crustal structure in Tunisia derived from gravity data. Geophysical Prospecting. 2015:1-12. DOI: 10.1111/1365-2478.12248

[46] Hlaiem A. Halokinesis and structural evolution of the major features in eastern and southern Tunisian atlas. Tectonophysics. 1999;**306**:79-95

[47] Saïd A, Chardon D, Baby P, Ouali J. Active oblique ramp faulting in the southern Tunisian atlas. Tectonophysics. 2011a;**499**:178-189

[48] Saïd A, Baby P, Chardon D, Ouali J. Structure, paleogeographic inheritance, and deformation history of the southern atlas foreland fold and thrust belt of Tunisia. Tectonics. 2011b;**30**:TC6004. DOI: 10.1029/2011TC002862

[49] Chihi H, De Marsily G, Belayouni H, Yahyaoui H. Relationship between tectonic structures and hydrogeochemical compartmentalization in aquifers: Example of the " Jeffara de Medenine " system, Southeast Tunisia. Journal of Hydrology: Regional Studies. 2015;**4**(part B):410-430. DOI: 10.1016/j.ejrh.2015.07.004

[50] Soyer C, Tricart P. La crise aptienne en Tunisie central: approche paléostructurale aux confins de l'Atlas et de l'Axe Nord-Sud. Comptes Rendus de l'Académie des Sciences Paris. 1987;**305**(II):301-305

[51] Zouaghi T, Ferhi I, Bédir M, Ben Youssef M, Gasmi M, Inoubli MH. Analysis of cretaceous (Aptian) strata in Central Tunisia, using 2D seismic data and well logs. Journal of African Earth Sciences. 2011;**61**(2011):38-61. DOI: 10.1016/j.jafrearsci.2011.05.002

[52] Abdulbaset MA, Van Wees JD, van Dijk PM, Cloetingh SA. Tectonics and subsidence evolution of the Sirt Basin, Libya. AAPG Bulletin. 2008;**92**(8):993-1027

[53] Coward MP, Ries AC. Tectonic development of north African basins. From: Arthur TJ, Macgregor DS & Cameron NR. (eds). Petroleum Geology of Africa: New Themes and Developing Technologies. Geological Society, London: Special Publications; vol. 207. 2003. pp.61-83

[54] Jolivet L, Faccenna C, Agard P, Frizon de Lamotte D, Menant A, Sternai P, et al. Neo-Tethys geodynamics and mantle convection: From extension to compression in Africa and a conceptual

model for obduction. Canadian Journal of Earth Sciences. 2016;**53**:1190-1204. DOI: 10.1139/cjes-2015-0118

[55] Chirchi A, Trémolière P. Nouvelles données sur l'évolution structurale au Mésozoïque et au Cénozoïque de la Sardaigne et leurs implications géodynamiques dans le cadre méditerranéen. Comptes Rendus Académie Sciences Paris. 1984; 298, Série II(n°20):889-894

[56] Guiraud R, Maurin JC. Le rifting en Afrique au Crétacé inférieur; synthèse structurale, mise en évidence de deux phases dans la genèse des bassins, relations avec les ouvertures océaniques péri-africaines. Bulletin Société Géologique France. 1991;**5**:811-823

[57] Guiraud R, Bellion Y. Late Carbonifereous to Recent Geodynamic Evolution of the West Gondwanian Cratonic Tethyan Margins. New York: The Tethys Ocean Plenum; 1995. pp. 101-124

[58] Ylles Chaouch AK, Ait Ouali R, Bracène R, Derder MEM, Djellit H. Chronologie de l'ouverture du bassin de Ksour (Atlas Saharien, Algérie) au début du Mésozoïque. Bulletin Société Géologique France, t. 2001;172(n°2): 285-293

[59] Ait Brahim L, Chotin P, Hinaj S, Abdelouafi A, El Adraoui A, Nakcha C, et al. Paleostress evolution in the Moroccan African margin from Triassic to present. Tectonophysics. 2002;**357**:187-205

[60] Platt JP, Behr W, Johanesen K, Williams JR. The Betic-Rif Arc and Its Orogenic Hinterland: A Review. Annual Review of Earth and Planetary Sciences. 2013;**41**:14.1-14.45. DOI: 10.1146/annurev-earth-050212-123951

[61] Laridhi Ouazaa N, Bedir M. Les migrations tectono-magmatiques du Trias au Miocènes sur la marge orientale de la Tunisie. Africa Geosciences Review. 2004;11(N°3):179-196

[62] Capitanio FA, Goes S. Mesozoic spreading kinematics: Consequences for Cenozoic central and Western Mediterranean subduction. Geophysical Journal International. 2006;**165**(3):804-816. DOI: 10.1111/j.1365-246X.2006.02892.x

[63] Rosenbaum G, Lister GS, Duboz C. Relative motions of Africa, Iberia and Europe during alpine orogeny. Tectonophysics. 2002;**359**(1-2):117-129. DOI: 10.1016/S0040-1951(02)00442-0

[64] Dercourt J, Zonenshain LP, Ricou LE, Kazmin V, Le Pichon X, Knipper L, et al. Présentation de 9 cartes paléogéographiques au 1/20000000 s'étendant de l'Atlantique au Pamir pour la période du Lias à l'Actuel. Bulletin Société Géologique France, 8. 1985;**I**(5):637-652

[65] Stampfli GM, Borel GD. A plate tectonic model for the Paleozoic and Mesozoic constrained by dynamic plate boundaries and restored synthetic oceanic isochrons. Earth and Planetary Science Letters. 2002;**196**:17-33

[66] Gong Z, Langereis CG, Mullender TAT. The rotation of Iberia during the Aptian and the opening of the Bay of Biscay. Earth and Planetary Science Letters. 2008;**273**:80-93. DOI: 10.1016/j.epsl.2008.06.016

[67] Vissers RLM, Meijer PT. Iberian plate kinematics and alpine collision in the Pyrenees. Earth-Science Reviews. 2012;**114**(2012):61-83. DOI: 10.1016/j.earscirev.2012.05.001

[68] Berra F, Angiolini L. The evolution of the Tethys region throughout the Phanerozoic: A brief tectonic

reconstruction. In: Marlow L, Kendall C, Yose L, editors. Petroleum Systems of the Tethyan Region: AAPG Memoir. 2014;**106**:1-27. DOI: 10.1306/13431840M1063606

[69] Menant A, Jolivet L, Vrielynck B. Kinematic reconstructions and magmatic evolution illuminating crustal and mantle dynamics of the eastern Mediterranean region since the late cretaceous. Tectonophysics. 2016;**675**:103-140. DOI: 10.1016/j.tecto.2016.03.00

[70] Ye J, Chardon D, Rouby D, Guillocheau F, Dall Asta M, Ferry JN, et al. Paleogeographic and structural evolution of northwestern Africa and its Atlantic marginssince the early Mesozoic. Geosphere, Geological Society of America. 2017;**13**(4):1254-1284. DOI: 10.1130/GES01426.1 insu-01533455

[71] Guerrera F, Martín-Martín M, Tramontana M. Evolutionary geological models of the central-western peri-Mediterranean chains: A review. International Geology Review. 2019;**63**(1):65-86. DOI: 10.1080/00206814.2019.1706056

[72] Agard P, Jolivet L, Vrielynck B, Burov E, Monie P. Plate acceleration: The obduction trigger ? Earth and Planetary Science Letters. 2007;**258**:428-441

[73] Carminati E, Doglioni C. Mediterranean tectonics. Encyclopedia of Geology, Elsevier. 2004;**1**:135-146

[74] Doglioni C, Gueguen E, Harabaglia P, Mongelli F. On the origin of W-directed subduction zones and applications to the western Mediterranean. Geological Society - Special Publications. 1999;**156**:541-561

[75] Lustrino M, Wilson M. The circum-Mediterranean anorogenic Cenozoic igneous province. Earth-Science Reviews. 2007;**81**(1-2):1-65. DOI: doi.org/10.1016/j.earscirev.2006.09.002

[76] Agard P, Zuo X, Funiciello F, Bellahsen N, Faccenna C, Savva D. Obduction: Why, how and where. Clues from analog models. Earth and Planetary Science Letters. 2014;**393**:132-145

[77] Hippolyte JC, Müller C, Kaymakci N, Sangu E. Dating of the Black Sea Basin: New nannoplankton ages from its inverted margin in the central Pontides (Turkey). In: Sosson M, Kaymakci N, Stephenson R, Bergerat F, Starostenko V, editors. Sedimentary Basin Tectonics from the Black Sea and Caucasus to the Arabian Platform. London: Geological Society; 2010. pp. 113-136. DOI: 110.1144/sp1340.1147

[78] Nishikin AM, Cloething S, Brunet MF, Stephenson RA, Bolotov SN, Ershov AV. Scythian platform, Caucasus and Black Sea region: Mesozoic-Cenozoic tectonic history and dynamics. In: Crasquin-Soleau S, Barrier E, editors. Peri-Tethys Memoir 3: Stratigraphy and Evolution of Teri-Tethyan Platforms. NameMémoires du Muséum National d'Histoire Naturelle. No. 177. Paris; 1998. pp. 163-176

[79] Nikishin AM, Okay A, Tüysüz O, Demirer A, Amelin N, Petrov E. The Black Sea basins structure and history: New model based on new deep penetration regional seismic data. Part 1: Basins structure and fill. Marine and Petroleum Geology. 2015a;**59**:638-655. DOI: 10.1016/j.marpetgeo.2014.1008.1017

[80] Nikishin AM, Okay A, Tüysüz O, Demirer A, Amelin N, Petrov E. The Black Sea basins structure and history: New model based on new deeppenetration regional seismic data. Part 2: Tectonic history and paleogeography. 811 marine pet. Geology.

2015b;**59**:656-670. DOI: 10.1016/j.marpetgeo.2014.1008.1018

[81] Kraiem Z, Zouari K, Bencheikh N, Agoun A, Abidi B. Processus de minéralisation de la nappe du Plio-QuaternairedanslaplainedeSegui-Zograta (Sud-Ouest tunisien). Hydrological Sciences Journal. 2015;**60**(3):1-15. DOI: 10.1080/02626667.2013.877587

Section 2

Geochemistry and Mineralization of the Earth's Crust

Chapter 4

Characteristic of Specialized Geochemistry of the Ngoc Tu Granitoid Block KonTum, Vietnam

Nguyen Duc Do, Niem Van Nguyen, Hieu Cong Duong, Tan Trong Bui, Linh Thuy Thi Hoang and Tien Cong Dinh

Abstract

The determination of magmatic mineralogy based on the oxidation-reduction state of magma by primary inclusion geochemistry of RAMAN equipment has identified Ngoc Tu granitoid blocks in the oxidation state represented by CO_2-rich primary inclusions. Redox conditions help explain the specialized chemistry of geochemistry not only at the content level, but also in terms of the geochemical behavior of the elements. This shows that the Ngoc Tu granitoid block is not favorable for the metal potential of Sn, but is favorable for the movement of Mo and W from the magma solution to the ore solution.

Keywords: Ngoc Tu granitoid block, oxidation-reduction, specialized chemistry of geochemistry, RAMAN, mineralogy

1. Introduction

Ngoc Tu block granite in particular has been studied by many works and has conception related to tin (Sn), wolfram (W) mineralization potential. However, there has not been any work that clearly confirms the Sn mineral-forming ability in Ngoc Tu block, although Duong Duc Kiem [1] discovered Sn mineralized expression. This result is based on the geochemical parameters of the epigenetic zone, with the geochemical document of the primary rock using semi-quantitative methods, which has no meaning in determining the specialization of magma. At the same time, the types of changes related to mineralization have not been detailed.

Meanwhile, the geochemical specialization of molybdenum (Mo) has not been determined; Sn, W, Cu are determined by semi-quantitative method. In particular, the oxidation-reduction state of a granitoid has not been studied, and this is a decisive condition for the geochemical behavior of Sn, Mo, W, Cu, ... capable of moving from the magma when bound.

To clarify this problem, surveyed and sampled (**Figure 1**) studied the system from the edge to the center of the mass, the facies of the granite, the variable zone, and related geological structure. Since then, building the quantitative database includes: (1) analysis

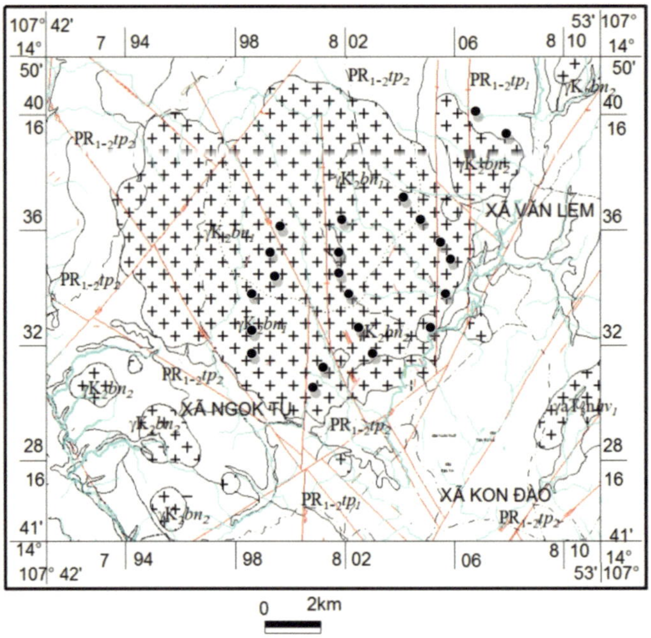

Figure 1.
Sampling location of Ngoc Tu granitoid block, KonTum, Vietnam (• Sampling point).

of 24 samples of thin sliced granite (12 samples of porphyr granite, 6 samples of small grain granite, 13 samples of modified granite rocks); (2) analysis of 21 geochemical samples of rare elements; and (3) analysis of protozoa composition in total rock.

2. Geological characteristics of Ngoc Tu block

The Ngoc Tu granite block (**Figure 2**) is one of the highest mountain peaks in the region, about 16 km northwest of Dak To district, KonTum province. Previously, Ngoc Tu granite block was established by Tran Tinh, Nguyen Van Trang, (1994) [2], Nguyen Quang Loc (1998) [3] and classified into Ba Na complex with age K2. Ngoc Tu block has a fairly isometric shape, an area of about 120 km^2. This granite block penetrates the biotite gneiss, plagioclase of the Tac Po formation, and the biotite granite of the Hai Van complex. Albitization and horny rocks are common at the contact edge: quartz horn—feldspar—mica, quartz horn—biotite—cordierite. The lithological composition of the block consists of brightly colored granite but has many different facies and the boundary is not clear, including: porphyritic to medium-grain granitoid, and small-grain granitoid. In addition, there is a vascular phase in the internal magma block (**Figure 3**).

- Phase 1 has a medium, relatively light-colored granite granite type with block structure. Ingredients mainly include feldspar, quartz, muscovite, biotite.

Mineral composition: The main minerals are plagioclase (40 ÷ 42%), potassium feldspar (orthoclass + microcline) 32 ÷ 34%, quartz 23 ÷ 25%, biotite 2%, muscovite from 1 ÷ 2%. Biotite is chlorinated.

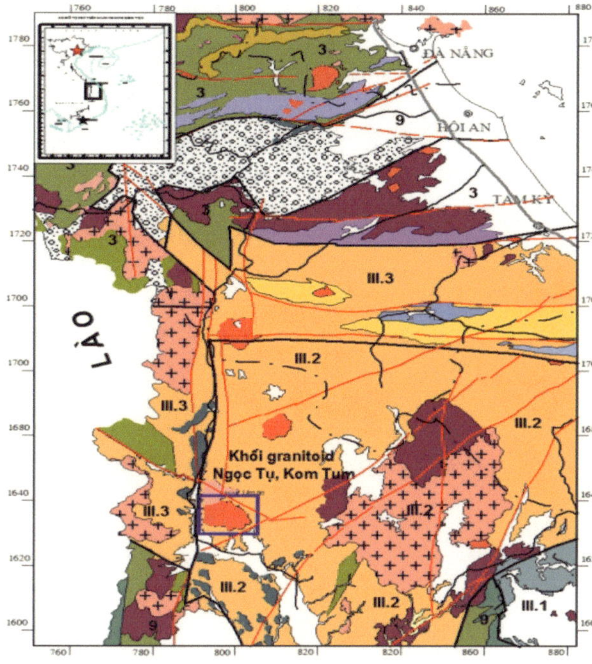

Figure 2.
Location of Ngoc Tu granitoid block in the Central Highlands region, Vietnam.

Figure 3.
Image of granitoid porphyr and granitoid medium-small.

- Phase 2 has a pattern of granite with small to medium grain, relatively dark color, with block structure. Ingredients mainly include feldspar, quartz, muscovite, and biotite.

Mineral composition: The main minerals are plagioclase (20 ÷ 31%), potassium feldspar (orthoclass + microcline) 34 ÷ 49%, quartz 26 ÷ 30%, biotite 1 ÷ 2%, muscovite 1 ÷ 2%. Biotite is often chlorinated.

- The berezitized change zone has the following characteristics: 1/The residual base rock is potassium feldspar (52 ÷ 53%) in a large slab form, very uneven distribution, opaque surface, along the open sand of potassium feldspar densely developed squamous sericite. 2/Hydrothermal part: very uneven distribution; the composition is quartz (35 ÷ 37%), distorted large grains, muscovite (3 ÷ 4%) in slabs, distorted, colorless, highly interfering; sericite (7 ÷ 9%) scaly, clean surface, often clumping—uneven drive.

The age of zircon formation in granitoid determined by U-Pb, zircon, SHRIMP method at AP Kapinski (VSEGEI) was 244.5 ± 1.5 million (T2) according to Nguyen Van Niem [4]. This result is similar to the determination of the isotopic age by U - Pb method for granite of Ba Na type in Ngoc Tu block for the age of 239 ÷ 240 million years, corresponding to T2 [5].

Tectonic faults: The northwest-southeast fault system has quite strongly divided, creating strong fracture zones, favorable for later hydrothermal processes.

3. Materials and methods

With the aim of clarifying the mineralogical role of the Ngoc Tu block granite formation, samples were collected for each rock type, zone of change, and mineralized expression zone, including lithological thin slices, basal geochemical samples and zone of change, and sample for study of the composition of primary magma.

The trace element geochemical samples were analyzed by ICP-MS method at the Far East Geological Institute-FEGI (Russia) systematically from the edge of the block to the center, characteristic of rock types including: a granite rich in pegmatite, medium grain granitoid, small grain granitoid, zone of change, zone of fracture tectonic, sulfide zone. This sample serves to study geochemical specialization, to determine minimum anomaly and mutation values that contribute to mineral search.

The protoplasmic specimen in magma is done in the following steps: Research and sampling ensure the representativeness of the sample for each rock type. Processing and analyzing two types of rocks including: small and medium grain biotite granite with transition boundary; small grain biotite rock in the tectonic break zone. Samples are processed to a thickness of about 1.0÷1.3 mm, polished, size 2.5×4.5 cm. The samples were processed and analyzed on Raman equipment at the Laboratory of Far East Geology (Russia). The protoplasmic holster is characterized by the following properties: It is contained in crystals and is part of a random connection; an indication of the direction of crystal growth can be observed if the capsule is large enough 1/10 times the crystals that contain them) or the form evolving in three directions (this is a very reliable sign); it can be in the center, edge, or crystal position of driftwood; in a crystal contains only 1 capsule; protozoa often have irregular, small (secondary or combined) envelopes arranged around or opposite.

The general analytical procedure is as follows: microscopy and protozoa detection, selection of suitable size capsule (> 1 pm, optimal 10 pm or more), marking position on slice, drawing preliminary shape—characteristics of the capsule; transfer to Raman device, prepare the device with sample C, put the sample of the specimen marked with the object into the analysis position, choose the full wavelength for scanning, separate each wavelength according to analytical composition, record and save data, and transfer it to the server system to process results. This method helps to study the primary inclusion composition of the aggregate rock, interpret the oxidation-reduction state of the granite most accurately, and evaluate the biochemistry of the granitoid. However, the microscopic rock type has not been able to detect the protozoa because the probability of detecting it in the rock is very low (about 10%) and their size is small (<1 pm). Only protozoa in rocks are used to study the oxidation-reduction state of magma.

In addition, the authors also collected, studied, and used the main components and traces of the rock to calculate the oxidation-reduction conditions. From there, compare with the primary holistic research method.

4. Results and discussion

4.1 Results of the analysis of the body

Fluid composition in primary inclusions of light-colored porphyry and medium-small-grained granites is analyzed by Raman instrument at the Institute of Far Eastern Geology, Sub-Institute of the Academy of Sciences of the Russian Federation

- Bright porphyr granite meets primary inclusions in quartz mineral background (**Figure 4**).
- Light porphyric granite transition from the first type rock (**Figure 4**) in this zone in the primary inclusion in quartz mineral.
- Light-colored medium-small-grained granite: This type of stone has a transition boundary.

In contact with light-colored porphyric granite, a primary inclusion was discovered.

The results of the analysis are as follows:

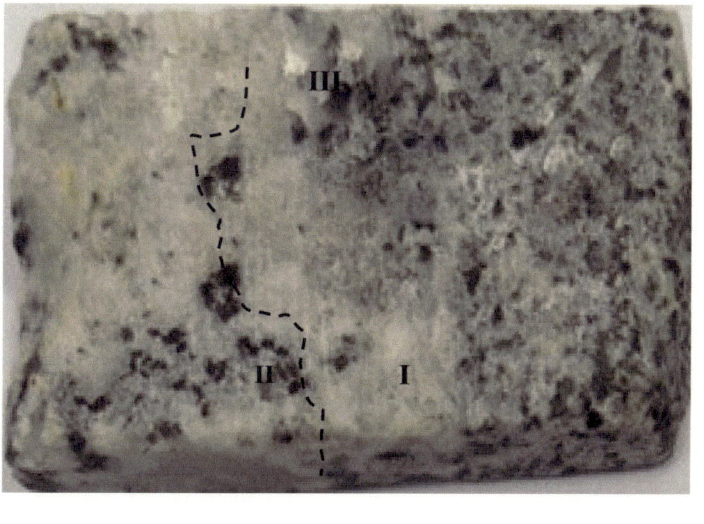

Figure 4.
Variation of granitoid porphyr (II, III) and granitoid medium-small (III) in Ngoc Tu granitoid block.

- Raman spectroscopy to determine the primary inclusion composition in CO_2-containing quartz minerals with a change in the granitoid from type 1 located in this zone, the primary inclusion in the quartz mineral containing CO_2 with density 0.23–0.35 g/cm^3 from 1283.7 to 1387.0 cm^{-1} to type 2, located in this rock type, distributed independently, far from the boundary of contact with metamorphic rocks as well. Primary inclusions are encountered in quartz minerals that exist with density 0.15–0.25 g/cm^3 with a spectral range of CO_2 from 1285 to 1388.1 cm^{-1} and type of 3 primary inclusion sites located at in quartz minerals exist with density 0.17–0.28 g/cm^3 with the spectral range of CO_2 from 1353 cm^{-1} (**Figure 5**).

- Raman spectroscopy determines the composition of primary inclusions in quartz minerals containing H_2O in liquid form of granitoid rock types, including: type 1 meets primary inclusions in the background of H_2O-containing quartz minerals with a spectral range of 3625 cm^{-1} in gaseous state to type 2 meets primary inclusions in H_2O-containing quartz minerals, the spectral range is 3575 cm^{-1} in liquid form, and type 3 meets primary inclusions in H_2O-containing quartz minerals with a spectral range of 3625 cm^{-1} in liquid form (**Figure 6**).

- Spectral band of primary inclusion of fluorapatite ($Ca_5(PO_4)3F$ surrounded by quartz crystals of granitoid rock types including: type 1 at the location where F—fluorapatite-rich primary inclusions ($Ca_5(PO_4)3F$) are encountered in quartz crystal has a spectral range: 464–963.2 cm^{-1}, type 2 at the location where F–fluorapatite-rich primary inclusions ($Ca_5(PO_4)3F$ in quartz crystals are found) has a spectral range: 429–465–965 cm^{-1} (**Figure 7**).

The above results show that, in Ngoc Tu block granitoid from the first rock type to the second and third type, there is an increase in the composition and concentration

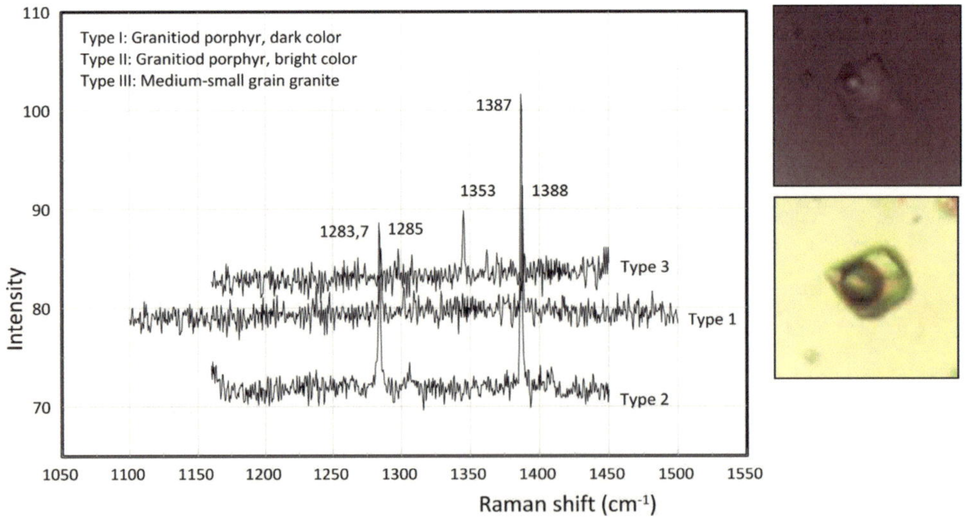

Figure 5.
Raman shift determines the primary inclusion composition in CO_2-containing quartz minerals of Ngoc Tu block granitoid types.

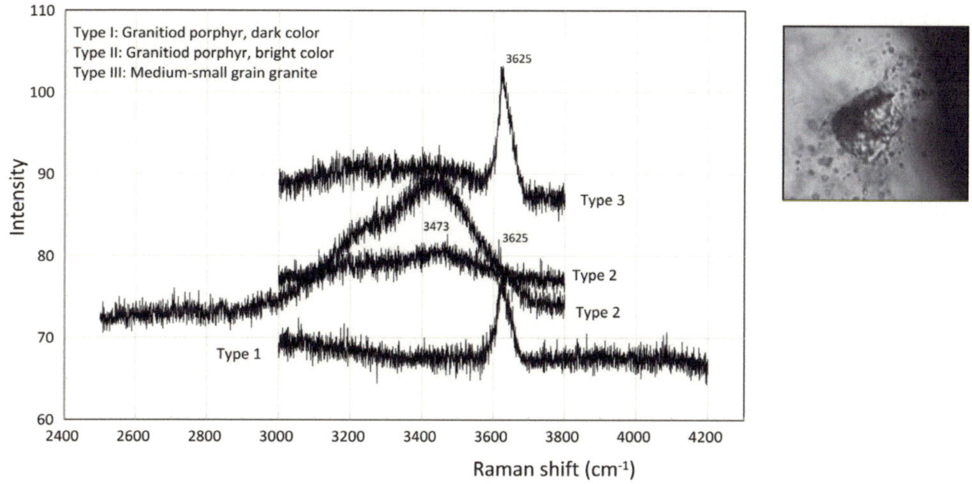

Figure 6.
Raman shift determines the primary inclusion composition in H_2O-containing quartz minerals of Ngoc Tu block granitoid types.

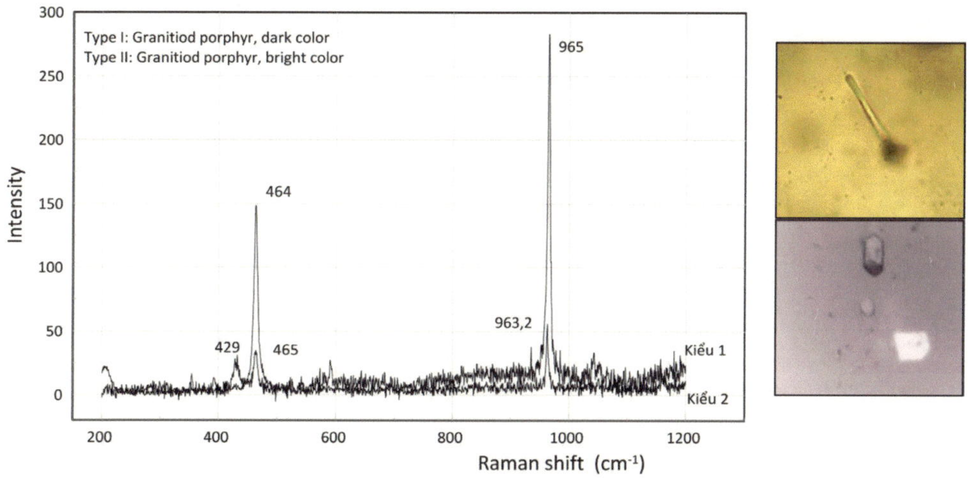

Figure 7.
Raman shift determines the primary inclusion composition in fluorapatite (Ca5(PO4)3F-containing quartz minerals of Ngoc Tu block granitoid types.

of fluid in the primary inclusion, specifically rich in CO_2 and H_2O, which is shown by the results, clearly oxidizing properties of rocks [6]. In addition, the appearance of F-rich inclusions is typical for the volatile composition and the small-medium-grained granite that was born after the expression of increased H_2O composition in the primary inclusion.

Thus, the research results on the magma environment show that the Ngoc Tu block granitoid has a moderate to strong oxidizing environment. This environment is typical for the mineralization of Mo, W, and U, which is not favorable for the formation of Sn deposits. Only two protoplasmic inclusions were found in a medium-grain biotite granitoid sample, which was contained in a quartz crystal.

4.2 Oxidation state of Ngoc Tu granitoid

Light-colored granite porphyr occupies the main area of Ngoc Tu mass. The analysis results show that the crystals in quartz crystals find CO_2 (**Figure 3**), which characterizes the primary, oxidizing magma environment.

Redox states are also expressed by the main components (Fe) and trace. However, these factors are very easy to be affected by the process of secondary transformation. Here, bright granite with $Fe_2O_3/FeO > 0.5$ ratio is typical for oxidized granitoid and vice versa, and bright granite with Fe_2O_3/FeO ratio < 0.5 is typical for reduced granitoid. For Ngoc Tu, according to the distribution and correlation characteristics between Fe^{2+} and Fe^{3+} [3], magma granitoid solids exhibiting moderate-to-strong oxidation (**Figure 8**) have favorable ability to generate W, W-Mo (**Figure 9**). However,

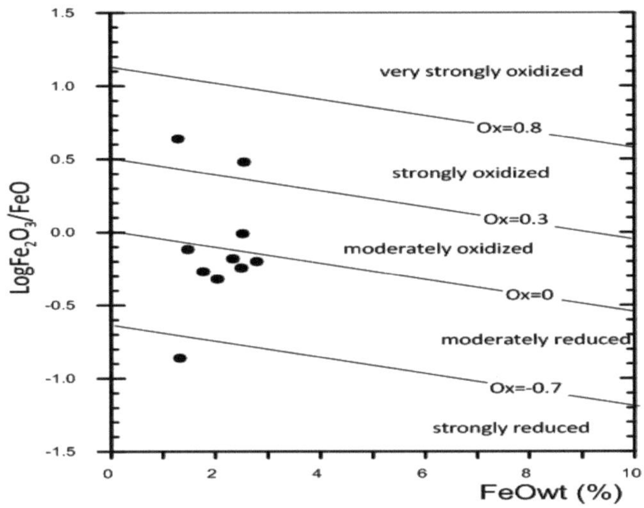

Figure 8.
Diagram of classification of oxidation-reduction conditions of Ngoc Tu block granite [7].

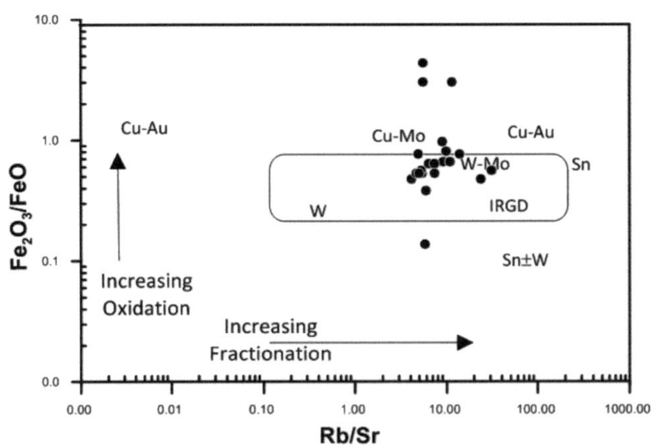

Figure 9.
Correlation diagram between oxidation conditions and differentiation level of Ngoc Tu granite in relation to mineralization [7].

the use of Fe^{2+} and Fe^{3+} to divide the geochemical environment of the jade granite magma also showed both oxidation and reduction (medium to strong) medium (**Figure 8**), which may be caused by granite rocks. This block is affected by chloritization, less muscovitization, and epidotization.

4.3 The geochemical specialization of the jadeite block Ngoc Tu

For large granitoid porphyria, there is no localization of Mo (Ktt = 1.36) Sn, W, Pb, Th are specialized. U has medium specialization, while Re shows very high specialization (**Table 1**). This result for Mo is in contrast to previous studies, and the reliability of the data is more assured. However, W is only for reference because the analysis has a large random error, but the analytical results of this element still clearly distinguish from the reality between the bedrock-change zone and mineralization zone.

NT	Min (ppm)	Max (ppm)	TB (ppm)	(S)	V (%)	Ktt	A	B
Li	16.17	117.98	59.88	30.27	50.54	0.75	80	
Be	3.14	9.81	5.07	1.84	36.22	1.45	3.5	
Sc	3.70	5.40	4.48	0.49	11.01	0.64	7	
V	10.49	23.36	16.45	3.28	19.92	0.23	70	
Cu	1.85	30.26	6.44	7.75	120.39	0.26	25	
Zn	21.02	37.38	30.46	4.91	16.12	0.53	58	
Ga	14.10	18.40	16.09	0.95	5.93	0.85	19	
Ge	2.68	3.91	3.01	0.30	9.86	2.01	1.5	
Rb	350.57	446.43	392.17	29.64	7.56	1.96		200
Sr	38.01	76.04	58.39	11.48	19.67	0.22	270	
Y	17.88	56.49	27.37	10.42	38.07	0.68	40	
Nb	13.29	19.88	15.77	1.83	11.61	0.79	20	
Mo	0.45	9.41	2.04	2.69	131.63	1.36	1.5	1
Cd	0.03	0.09	0.05	0.01	30.72	0.27	0.17	
Sn	5.15	10.81	7.30	1.54	21.06	2.43	3	
Cs	11.62	32.78	20.33	6.33	31.13	4.07		5
Ba	200.54	538.59	328.30	84.62	25.77	0.47	700	
Hf	2.44	4.40	3.56	0.61	17.17	3.56		1
Ta	1.63	3.61	2.55	0.66	25.83	1.02	2.5	
W	0.90	44.50	6.57	10.63	161.69	**3.29**	2	
Re	0.00	0.02	0.01	0.00	33.79	18.78	0.00067	Re
Pb	35.40	48.51	43.97	3.63	8.25	**2.20**	20	
Th	27.93	52.35	38.35	6.97	18.18	**2.13**		18
U	4.44	32.52	12.35	7.79	63.02	**3.53**		3.5

Table 1.
The geochemical parameters of the Ngoc Tu block porphyritic granitoid (n = 12 samples).

NT	Min (ppm)	Max (ppm)	TB (ppm)	(S)	V (%)	Ktt	A	B
Li	22.09	35.03	27.87	5.34	19.16	0.35	80	
Be	3.03	13.32	7.42	4.40	59.35	2.12	3.5	
Cu	5.98	28.10	20.15	10.28	50.99	0.81	25	
Zn	8.53	15.92	12.76	3.08	24.14	0.22	58	
Ga	13.64	19.81	15.66	2.82	17.99	0.82	19	
Ge	2.10	2.37	2.21	0.12	5.40	1.48	1.5	
Rb	399.28	442.14	418.95	17.62	4.21	2.09		200
Sr	16.51	38.45	28.08	9.20	32.77	0.10	270	
Y	8.67	18.51	14.82	4.48	30.23	0.37	40	
Zr	26.22	53.33	36.46	11.75	32.22	0.18		200
Nb	8.75	15.72	12.42	3.11	25.00	0.62	20	
Mo	2.84	11.10	6.76	4.35	64.34	4.51	1.5	
Cd	0.03	0.03	0.03	0.00	11.63	0.17	0.17	
Sn	4.46	8.17	5.89	1.69	28.67	1.96	3	
Ba	29.26	68.83	56.33	18.23	32.36	0.08	700	
Hf	1.28	1.71	1.44	0.19	13.43	1.44		1
Ta	1.27	3.98	2.32	1.19	51.43	0.93	2.5	
W	2.76	15.78	9.27	5.69	61.43	4.63	2	
Re	0.00	0.01	0.01	0.00	40.42	13.22	0.00067	Re
Pb	37.42	59.31	49.22	9.98	20.28	2.46	20	
Th	9.35	13.51	11.38	1.76	15.43	0.63		18
U	10.79	29.36	21.23	8.42	39.65	6.07		3.5

Note: Value of clark of elements in granite: A— according to A.A. Golovin (2000); B—according to A. P. Vinogradov (1962).

Table 2.
The geochemical parameters of the Ngoc Tu small grain granitoid (n = 6 samples).

For the granitoid of small particles, element Mo has high specialization (Ktt = 4.51) (**Table 2**), close to the high geochemical threshold. This result is determined for the first time by the same spatial and time relationship with the large-grained porphyritic granite and quartz-molybdenite ore circuits (the results of previous studies have not clarified this). The set of datasets are tested to ensure calculation. Other elements such as W have high geochemical specialization, U has a geochemistry specialization (Ktt = 6.07), Pb also has geochemical specialization, and Sn also exhibits geochemical specialization (**Table 2**) but is low than porphyr granite. Thus, Mo, U, W in small granitic granite have much higher specialized properties than large grain porphyritic granite.

4.4 Discussion

There are some differences in petrographic characteristics with previous research results: porphyr granite. Small grain granite is rare. The common bedrock change phenomenon includes: chlorination, albitization, epidotization, bezeritization; greisen turned. These characteristics form a solid basis for the interpretation of geochemical parameters and geochemical behavior of elements when determining localization.

The results on the protozoa composition characteristic for Ngoc Tu block-shaped granite magma medium have oxidation-reduction properties. Thus, in the evolutionary process of Ngoc Tu block magma from porphyritic granite with small granites increasing the composition and concentration of fluid in the primitive, rich in CO_2, H_2O clearly shows oxidation [6, 8] of granite study area, in which the sample did not detect any type of fluid. In addition, the appearance of F-rich capsules characterizes the volatile composition. Postpartum granitoid showed an increase in H_2O content in protozoa.

In addition to the primary mitochondria, there are also mixed and secondary mitochondria. Banded distribution characteristics, according to cracks or cracked protoplasm forming a foam form of small particles, indicate that the area is deformed. This phenomenon is probably also consistent with the process of cutting through the ancient metamorphic formations in the area that changes the specific pressure of the steam within the contact range, forming cracks in the internal mass and the redistribution of material components, including Mo.

The average arithmetic average content of Mo in Ngoc Tu mass is very high (135.94 ppm). Meanwhile, the weighted average content typical for Ngoc Tu block granite is 1.52 ppm (this value is usually lower than the value of the local background content). This very large difference between the arithmetic mean and this weighted mean reflects the very difference between the minimum negative anomaly (0.45 ppm) and the maximum content-max (3134.34 ppm) of sample set (18 samples) (**Table 3**, **Figure 10**). At the same time, this maximum value and some high

Parameter/ element	**Unit**	**W**	**Cu**	**Sn**	**Mo**	**Pb**	**Zn**	**U**
No. of samples		18	18	18	18	18	18	18
% value of mutation	%	11.54	7.69	7.69	7.69	11.54	3.85	0.00
Standard deviation		296.86	114.42	6.90	625.78	27.94	13.52	296.86
Variation coefficient (V)	%	333.53	277.11	79.16	460.32	49.84	51.42	62.38
Min	ppm	0.90	1.85	1.41	0.45	35.40	7.13	1.38
Max		1445.90	501.04	35.04	3134.31	169.78	72.13	32.52
Median		5.63	5.37	6.90	1.52	46.22	27.44	10.79
Trung bình		89.01	41.29	8.71	135.94	56.06	26.30	13.08
Minimal positive anomaly		21.00	40.20	12.81	12.33	71.35	60.04	32.52
Minimal negative anomaly		0.90	1.85	1.47	0.45	35.40	7.13	1.38

Parameter/ element	Unit	W	Cu	Sn	Mo	Pb	Zn	U
Number of mutation values		2	2	0	2	0	0	0
Number of values above the minimum positive anomaly threshold		3	2	1	2	3	1	0
Number of values below the minimum negative anomaly threshold		0	0	1	0	0	0	0
Maximum value of mutation	ppm	1445.90	501.04	35.04	3134.31	169.78	72.13	
Minimum value of mutation				1.41				
Contrast of anomaly		68.85	12.46	2.73	254.21	2.38	1.20	

Notes: Minimum positive anomaly is a value indicating the ability to be related to the accumulation of an element that can form minerals. Minimum negative anomaly is a value that represents the possibility that it is involved in the abnormal movement of an element, which can lead to ore formation in a different location favorable in terms of geographic conditions.

Table 3.
The geochemical parameters of Mo and related elements of Ngoc Tu granitoid.

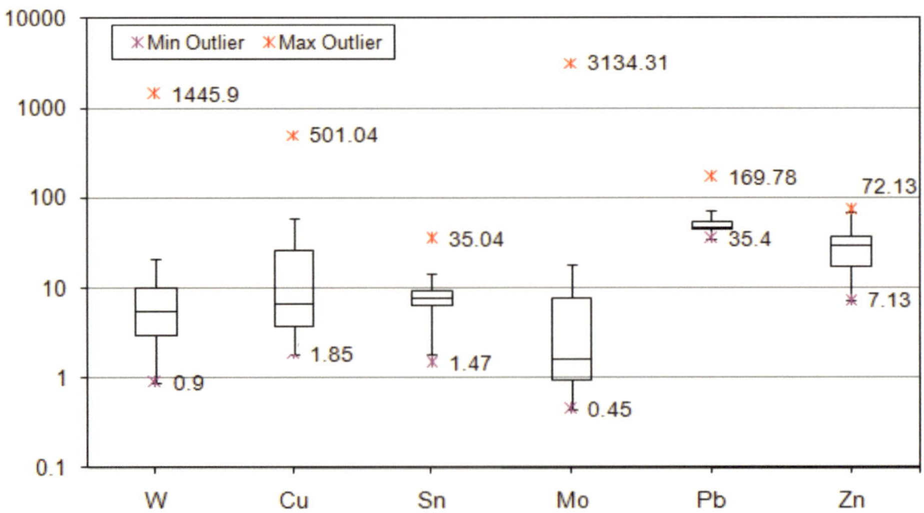

Figure 10.
Diagram of determining the geochemical parameters (minimum anomaly, weighted average, mutation value ...) of Ngoc Tu granitoid (Note: the max outlier on the chart is the value maximum, which can be equal to the spike value).

content points (189.3 ppm) belonging to the max outlier sample were very significant for the search. Here, these two points (they are next to each other) have found quartz-molybdenum ores and sulfides of Cu, ... Along with these two Mo mutant samples are also two points with mutation in Cu content and W. At the same time, at the variable zone near the quartz-molybdenite ore circuit, the mineral uraninite was found but the U content was not high (<10 ppm).

- The minimum anomaly value of Mo is 12.33 ppm. Above this threshold of content, there is an anomaly of 14.27 ppm, this point belongs to the bezeritized zone, containing several molybdenite particles in the region containing quartz-molybdenite small vessels (large particles, rose form) and many agar microchips. he covers an area of several hundred meters.

- The coefficient of variation of Mo content is very high (V = 460.32%) and the anomalous contrast of Mo is very large (254.21 ppm). This is possible, the mobility of Mo from the bedrock to the environment with favorable geological conditions will accumulate and create minerals. W, Cu also have high coefficient of variation, but the contrast of anomaly is much lower than Mo (**Table 3**). The distribution characteristics of Mo (**Figure 11**) are very different in small granites and large grain porphyrites. Meanwhile, in the porphyritic granite zone, it contains trans-shear quartz-molybdenite veins (**Figure 12**).

Regarding the correlation characteristics between Mo and elements: Mo is an element that has no relationship (positively correlated) with any element in the Ngoc Tu block granite. At the same time, it is inversely correlated with the rare earth element group (REE), Sn, Zn, Ta, Hf, Th.

Figure 11.
Small particle molybdenite disseminated in Ngoc Tu granitoid microscopic fissure, sericiteized, less muscovitized, albitized.

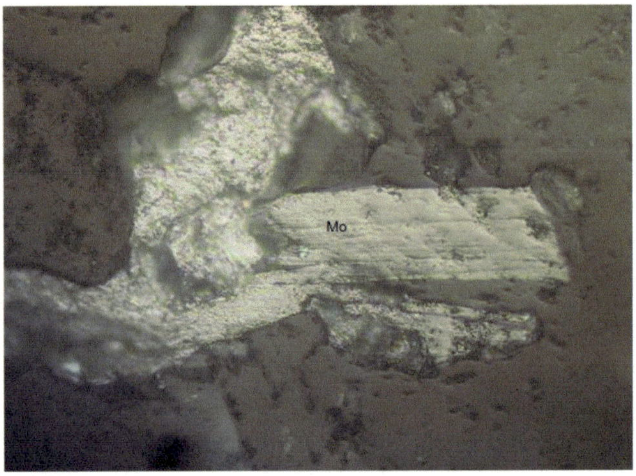

Figure 12.
Molybdenite dissemination in modified rock bed in Ngoc Tu granite. Nicol (−), 100×.

5. Conclusion

The properties of Ngoc Thu mass granitoid expressing both oxidation-reduction media based on the study of crystals in quartz crystals find CO_2 and the redox-redox state with main components (Fe) favorable for the mineralization of Mo, Cu–Mo, and W. Ngoc Tu block granitoid with high Re, Mo, W, U specialization, no mineralization of Sn. On the basis of the geochemical specialization Mo is quite suitable for accumulation such as the mineralization point Mo, W.

Author details

Nguyen Duc Do*, Niem Van Nguyen, Hieu Cong Duong, Tan Trong Bui, Linh Thuy Thi Hoang and Tien Cong Dinh
Vietnam Institute of Sciences and Mineral Resources, Vietnam

*Address all correspondence to: nguyen180@gmail.com

References

[1] Kiem DD. Research on mineralogy and forecast of mineral prospects in the Po Co tectonic zone. Hanoi: Geological Archives; 2006

[2] Tinh T, Van Trang N. Report on the results of geological measurement and mineral search in Kon Tum - Buon Me Thuot group of sheets at the scale of 1/200,000. Hanoi: Geological Archives; 1994

[3] Nguyen Quang Loc. Explanatory report, Geological and mineral map of Dak To group, scale 1: 50,000. Hanoi: Geological Archives; 1998

[4] Van Niem N, Nguyen DD. Study on Geochemistry of Molybdenum of Ba Na Granitoid Formations and their Mo Mineralization Potential. Hanoi: Institute of Geosciences and Minerals; 2018

[5] Tran Hoang Vu. U-Pb isotope age in zircon of Ngoc Tu block granite, Dak To district, Kon Tum province. In: Scientific Conference celebrates the 75th anniversary of the establishment of Geology. General Department of Geology and Minerals. Hanoi; 2015

[6] Moussallam Y, Edmonds M, Scaillet B, Peters N, Gennaro E, Sides I, et al. The impact of degassing on the oxidation state of basaltic magmas: a case study of Kīlauea volcano. Earth and Planetary Science Letters. 2016;**450**:317-325

[7] Blevin PL. Metallogeny of granite rocks. In: The Ishihara Symposium: Granites and Associated Metallogenesis. Vol. 14. Geoscience Australia; 2004. pp. 5-8 www.ga.gove.au

[8] Bakker RJ. Fluid inclusions as metamorphic process indicators in the Southern Aravalli Mountain Belt (India). Contributions to Mineralogy and Petrology. 2000;**139**(2):163-179

Section 3

Plate Tectonics and the Evolution of Biodiversity

Chapter 5

Perspective Chapter: Plate Tectonics and Global Dynamics of Biodiversity

Valeriy V. Snakin

Abstract

The article analyses the relationship between tectonic plate movements and global biodiversity dynamics. A new tectonic hypothesis of the causes of mass extinctions of biological species, which have repeatedly occurred in the history of the biosphere, is considered. It is proposed that the triggering mechanism for such extinctions is the periodic unification of lithospheric plates, leading to the formation of supercontinents and, consequently, to a decrease in the geographical isolation of species, increased interspecific competition, climate change, a decrease in the shelf area, changes in global sea level, and other determinants of biodiversity. The separation of lithospheric plates increases the degree of geographic isolation and, conversely, causes a new burst of species diversity that is greater than the previous one, as it engages the species best suited to the changing environment and wins the competition. The tectonic factor (geodynamics) thus becomes an important factor in the dynamics of biodiversity and evolution in general. Similarly, the dynamics of biodiversity are affected by contemporary processes of globalization: the human-accelerated reduction of habitats for native species, the removal of the factor of geographic isolation due to introduction, and increasing invasions are also leading to another mass extinction.

Keywords: biodiversity, mass extinction of species, geographical barriers, evolution of the biosphere, geodynamics, continental drift, competitive exclusion principle, globalization

1. Introduction

Elucidating the causes and mechanisms of mass extinctions of species, which have repeatedly occurred over more than 4 billion years of the biosphere's history, is one of the most interesting unresolved issues in the evolution of life on Earth. The number of species inhabiting the biosphere periodically dramatically decreased against the background of a general growth trend. There are more than a dozen hypotheses to explain this, which consider phenomena of a very different nature as the main cause, namely: cosmic catastrophes (first of all, the fall of large meteorites, other phenomena of cosmic origin); supervolcano eruptions; trap magmatism; climate changes (glacial cycles or even temperature increases critical for certain species [1]); changes in the level of the World Ocean; changes in the chemistry of the atmosphere and oceans (including changes in the atmospheric oxygen level); changes in Earth's magnetic field (periodic exchange of the south and north poles [2]); genetic problems (in particular, genetic diversity loss, diseases, such as osteoporosis in extinct mammoths [3]); and even the anthropogenic factor, despite its "biospheric youth." However, none of these hypotheses is universal; each one explains, as a rule, only one of the extinctions, considering coincidence in time only.

It should also be noted that the events considered as the root cause of mass extinctions occur much more often than mass extinctions themselves, and for some reason, their other manifestations are not associated with mass extinctions. For example, the death of dinosaurs at the border of the Mesozoic and Cenozoic is associated with the formation of one of the largest impact craters (astroblems) on Earth – Chiksulup – with a diameter of about 180 km on the Yucatan Peninsula. At the same time, the formation of a number of craters much larger than Chicksulup in size (namely, Warburten in Australia – 400 km, Vredefort in South Africa – 300 km, Sadbury in Canada – 250 km, and Dowling in China – 190 km) was not accompanied by subsequent extinctions and even global climatic changes, although they significantly exceeded the Yucatan event in terms of impact.

In our opinion, since mass extinctions have taken place in the history of the biosphere repeatedly, even with some regularity and in the same way, there must be one reason that prompts these biospheric crises.

2. Global dynamics of biodiversity

Table 1 summarizes information about the most famous mass extinctions according to the literature data with their supposed root cause.

Why, for what, and how do extinctions occur? There are still no unambiguous answers to these basic questions, which arise in the analysis of biodiversity dynamics.

The extinction of species as a result of natural processes is a normal phenomenon, balanced in geological time by the appearance of new species. According to C. Darwin [6], the extinction of species and their entire groups, which played such an outstanding role in the organic world history, is an almost inevitable consequence of the principle of natural selection. According to V.A. Krassilov [7], extinction is a way of regulating diversity under variable conditions, which weakens competition.

From the evolutionary viewpoint, the cause for the extinction of species is considered to be their permanent improvement in order to maximize the use of the environmental resources, which, in turn, is changed by new species in their favor. In the biosphere evolution process, the number of species increased, the biosphere spread to areas unoccupied by life, included new substances in the orbit of its activity, and utilized the energy of sunlight and chemical compounds more and more efficiently. Academician M.A. Fedonkin [8] noted: "extinctions were a boon for the biosphere, just as the death of an individual from old age is a boon for the species. In both cases, the carriers of inert hereditary information, which hinders evolution are cut off."

In the evolution process of the biosphere, the number of species was continuously increasing: from the Cambrian to the Silurian, over 200 million years, the number of families of marine animals increased from 5 up to 400. By the Quaternary period, the number of families of marine animals alone increased to 750. At the same time,

Major mass extinction (Size of the extinction)	When (million years ago, Mya)	Supposed causes of mass extinctions
The Ordovician-Silurian extinction mass extinction (~ 85% of all living species eliminated)	The Paleozoic Era, End-Ordovician (~ 440–445 Mya)	• Beginning of glacial cycles on Earth, and corresponding changes in sea level; • Changes in atmospheric and oceanic chemistry relating to the rise of the Appalachian mountains.
The Late Devonian mass extinction (~ 80% of all living species eliminated)	The Paleozoic Era, End-Devonian (~ 383–359 Mya)	• Climate change, possibly linked to the diversification of land plants; • Decrease in oxygen levels in the deep ocean.
The Permian mass extinction (~ 96% of all living species eliminated)	The Paleozoic Era, End-Permian (~ 250–252 Mya)	• Volcanic activity; • Climate change; • Decrease in oxygen levels in the deep ocean; • Changes in atmospheric chemistry; • Changes oceanic chemistry and circulation.
The Triassic-Jurassic mass extinction (~ 50% of all living species eliminated)	The Mesozoic Era, End-Triassic (~ 200 Mya)	• Volcanic activity
The Cretaceous-Paleogene (K-T) mass extinction (~75% of all living species eliminated)	The Mesozoic Era, End-Cretaceous (~ 66 Mya)	• Asteroid impact; • Volcanic activity; • Climate change; • Changes in atmospheric and oceanic chemistry.

Table 1.
Major mass extinction and geodynamic situation (by [4, 5]).

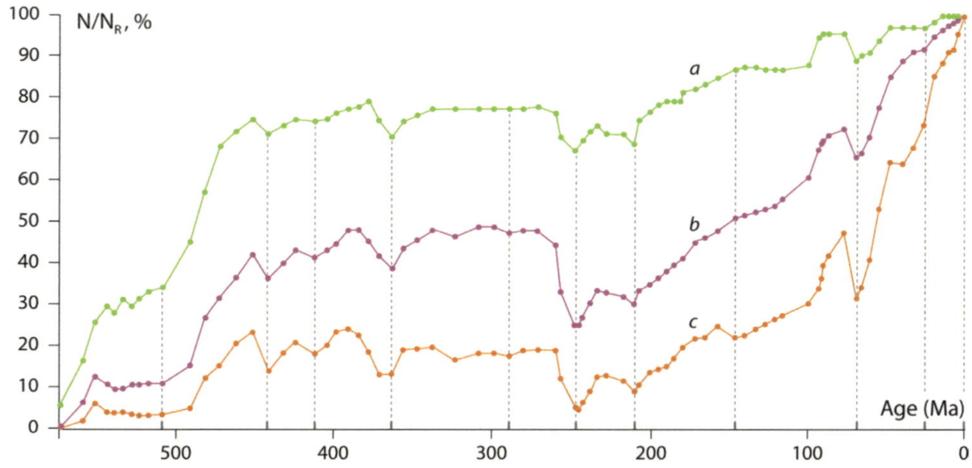

Figure 1.
Changes in the diversity of marine animals: A – Orders, b – Families, c – Genera. N is the number of taxa that existed at the lower boundary of the century, and N_R is the number of modern taxa known in the fossil state [9].

the dynamics of biodiversity, as well as the evolutionary process as a whole, was of a spasmodic nature: there were several large bursts of speciation and extinction, when entire families, orders, and even types of organisms appeared or disappeared. Several mass extinctions have been identified over the past 500 million years. The most famous, although not the largest, Cretaceous-Paleogene extinction (~65 million years ago) took away more than half of the species, including dinosaurs. During the period of the "great" Permian extinction (~250 million years ago), according to some estimates, ~90% of the species became extinct.

In the period accessible to the paleontological record, five mass extinctions are identified, as a rule; they are viewed in **Figure 1** and listed in **Table 1**. J.D. Corso et al. [10] proposed to add the extinction in the Carnian pluvial period (~233 Ma ago) to them as a sixth one. Ceballos et al. [11] proposed to consider the modern period associated with human activity as the sixth mass extinction.

3. Features of mass extinctions

An analysis of the nature of mass extinctions allows us to note some of their characteristic features, namely:

1. after a sharp decrease in the diversity of biological species during their death, at the next stage of evolution, biodiversity was restored and subsequently exceeded the precrisis level;

2. the duration of mass extinctions is sometimes estimated as millions of years (the Late Devonian extinction, according to Racki et al. [12], lasted ~13 million years), which does not correspond to the duration of catastrophic events and their consequences (as a rule, few years);

3. mass extinctions concerned various animal species to a much greater extent than plants, which cannot be explained within the framework of the existing, especially catastrophic, theories; and

4. most of the events indicated as the causes of mass extinctions for some reason acted as such only once, although they occurred regularly. For example, giant eruptions of supervolcanoes occurred quite regularly. For example, the Yellowstone supervolcano erupted relatively often: 640 thousand years, 1.3 and 2.1 million years ago.

Since mass extinctions have been repeated more than once in the history of the biosphere, there must be one root cause to prompt global fluctuations in the numbers of species.

Without downplaying the role of catastrophic events in the decline in the abundance of many biological species, let us try to look at mass extinctions from an evolutionary position, taking into account the horizontal and vertical movements of lithospheric plates. These movements lead to the consolidation and separation of continents, islands, and, accordingly, to changes in the geographical isolation of biological species. Some considerations in this regard were expressed by us earlier [13, 14]. A possible relationship between tectonics and the mass extinction was discussed in refs. [15, 16] when the formation of the Pangea supercontinent most obviously coincided with the largest Permian-Triassic extinction. At the same time, a somewhat belated extinction effect was noted.

4. Consolidation of lithospheric plates as the root cause of mass extinctions

The development of life on our planet over the billions of years of the biosphere's existence was under the continuous influence of geological processes, both catastrophic (volcanism, earthquakes, etc.) and much slower, but permanent in their manifestation, such as horizontal and vertical movements of Earth's crust. Such movements cause changes in climate, orography, and length of coastlines as well as changes in other environmental conditions, including the isolation factor of biological species.

As the main universal root cause of mass extinctions, we propose to consider relatively slow movements of lithospheric plates, due to which, periodic consolidations and separations of continents and islands occur and large continental associations are formed up to a single supercontinent such as Pangea (**Figure 2**), Gondwana, Rodinia, etc.

With this, the movements of lithospheric plates significantly change the *isolation factor*, which plays a very important role in speciation. Populations of species turn out to be either united over vast expanses or disunited on separate continents and islands.

It is important to note that the periodic consolidation and divergence of lithospheric plates had a regular character. **Figure 3** illustrates the periodic merging of lithospheric plates to form the supercontinents Pangaea, Gondwana, Rodinia, Columbia, etc.

Further movement of lithospheric plates, according to the hypothesis of the American geologist Ch. Scotese [21], will lead to the formation of a new single supercontinent Pangaea Ultima in 200–300 million years when the British Isles will turn to be in the North Pole region, while Alaska and Siberia will move to the subtropics.

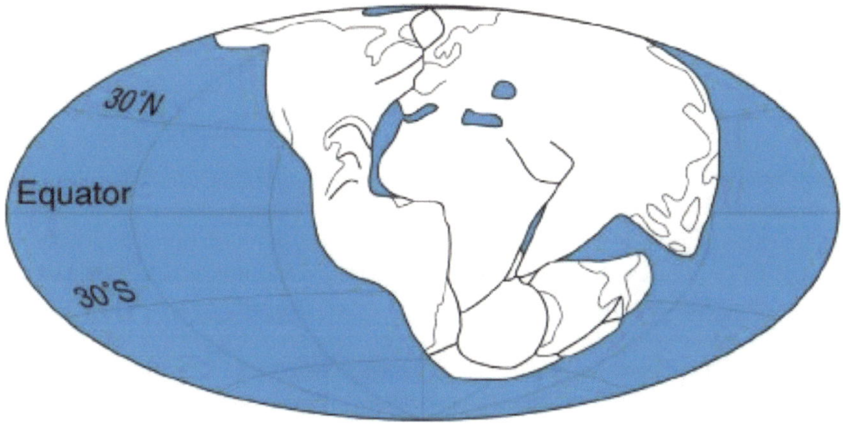

Figure 2.
The late Paleozoic supercontinent Pangea as envisioned by Wegener (1922) (from [17, 18]).

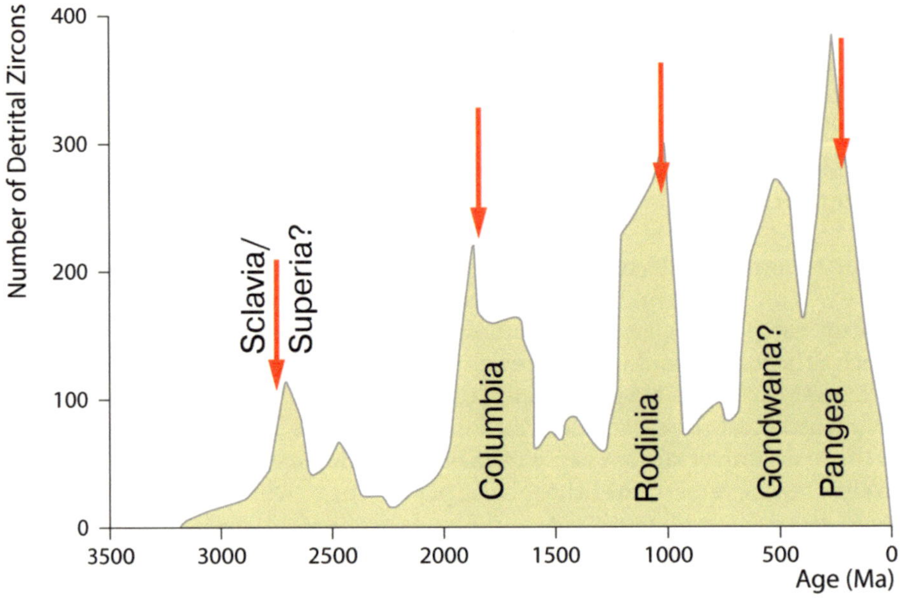

Figure 3.
Evidence of the association of lithospheric plates based on the comparison of orogenic peaks (arrows) recognized on radiometric data with spectra of U – Pb detrital zircon crystallization ages and proposed times of supercontinent assembly [19, 20].

Other versions of the next supercontinent are also being considered: the Novopangea, Amasia, Pangea Proxima, and Aurica [16].

Ecological consequences of the consolidation of lithospheric plates. In an ecological sense, the consolidation of lithospheric plates leads to the removal of geographical barriers and a decrease in the factor of geographic isolation of species.

When the continents unite, not only do the climatic and geomorphological[1] conditions change but also several species belonging to the same ecological niche in the united spaces may well merge. As a result, interspecific competition increases, and the competitive exclusion rule[2] comes into play.

According to this rule, two or more species cannot coexist stably within a limited space if they occupy the same ecological niche. The process of competition in this case always proceeds until the complete displacement of one species by another. Species less adapted to environmental conditions die out.

[1] Consolidations of continents due to their collisions are often associated with the activation of mountain building, which is the cause of the emergence of geographical barriers for biological species. However, this is a very slow process of movement (of the order of several cm per year), which is incomparable with the lifespan of species. Therefore, at the first stage of the consolidation of continental plates, animals have a lot of time to migrate to new territories.

[2] The competitive exclusion rule, or the Volterra-Gause principle, was formulated in 1926 by Vito Volterra based on the study of a mathematical model of the dynamics of two populations competing for one food resource. In 1931–1935 Georgy F. Gause, by experiments on protozoa, showed how competitive displacement of one species by another occurs. The discussion of the competitive exclusion rule played an important role in the development of the ecological niche concept and the ecological-geographical model of speciation, as well as in the assessment of interspecific competition as a factor to maintain the structure of communities [22].

Another important consequence of the unification of continents is the reduction in the number of ecological niches due to the decrease in the diversity of climate and other geographical conditions, which also reduces the opportunities for the existence of species. At the same time, the length of coastlines is reduced and the area of life-rich shelf ecosystems is sharply reduced.

Consequently, the consolidation of lithospheric plates, due to the amalgamation of species in similar ecological niches and the decrease in their diversity, triggers the mechanism of interspecific competition. In the further process of extinction of species, a variety of factors become important, namely: physiological characteristics, the degree of adaptation to food resources, and changed climatic and other geographical features.

For example, presumably, when six continental plates are combined, up to six different species could appear within one ecological niche, five of which are evolutionarily doomed to slow extinction due to strong interspecific competition. Perhaps that is why the most massive was the Permian extinction 250–252 Ma ago, which claimed up to 90% of species and was timed to coincide with the unification of most lithospheric plates into a single supercontinent Pangea (see **Figure 2**). But for a mass extinction, the union of 2–3 lithospheric plates is enough, which can threaten the extinction of up to half of the species.

Similar processes proceed in the oceanic expanses, where a single ocean (Panthalassa) is formed instead of several geographically isolated oceans and seas. At the same time, the area of the shelf zone saturated with life significantly reduces.

When the continents diverge, the surviving species in the new conditions of geographical isolation give rise to a new round of evolution to surpass the previous one, since the most evolutionarily developed species adapted to changing environmental conditions participate therein.

In the future, the convergence of the continents in new geographical conditions leads to a new cycle of extinction and a subsequent increase in biodiversity during their divergence (isolation).

Therefore, the consolidation of the continents and the accompanying extinction contribute to the preservation and further evolution of the most adapted species, which, in the course of the subsequent separation of the continents, gave rise to new even more promising species.

A comparison of the times of mass extinctions and collisions of lithospheric plates (**Table 2**) shows that almost all five identified mass extinctions occurred after significant mergers of lithospheric plates up to the formation of a single supercontinent. It is important to note that the process of extinction due to interspecific competition is a very slow process that occurs over many generations of species.

4.1 Vertical tectonic movements and biodiversity

Similarly to the horizontal movements of lithospheric plates, the ascending and descending movements of Earth's crust lead to transgression and regression of the seas, that is, the processes of amalgamation-separation of territories, and, accordingly, to changes in the factor of geographic isolation and biodiversity.

For example, the situation with the late Pleistocene extinction of megafauna in the period 20–4 thousand years ago (about 36% of large mammals died out in Northern Eurasia and about 72% did in North America) can be explained by the periodic formation of the Eurasia–America supercontinent due to the periodic formation and disappearance of the Bering Isthmus. It is known that over the past 3 million years, the territory of Beringia has risen and again submerged under water about 6 times [23]. The last time Eurasia and North America separated was 10–11 thousand years ago.

Major mass extinction	Associations of the continents
Ordovician-Silurian extinction mass extinction	Formation of Laurussia (Euramerica), the Paleozoic supercontinent, as a result of the collision of the North American platform (the ancient continent of Laurentia) and the East European platform (the ancient continent of Baltica) during the Caledonian orogeny (500–400 Mya).
Late Devonian mass extinction	In the Devonian, the northern continents formed a single large continent (Atlantia), to the east of which was Asia. Gondwana continued to exist.
Permian-Triassic mass extinction	Pangaea formed in the second half of the Carboniferous and Permian (~ 300–180 Mya). This period was long and took place in several stages.
Triassic-Jurassic mass extinction	Collision of the continents broke away from Gondwana with parts of Laurasia.
Cretaceous-Paleogene (K-T) mass extinction	In the Paleogene (~ 65 Mya), the territory of Eurasia increased: the Sundian archipelago joined the mainland, the Balkan Peninsula formed a single whole with Asia Minor, Europe joined Africa in the area of modern Gibraltar, and there was a merging with the North America in the northwest.

Table 2.
Consolidations of lithospheric plates during mass extinctions.

Before this, the isthmus had existed for 15–18 thousand years. Of the large herbivores, the mammoth and wooly rhinoceros became extinct at this time; of the large predators, giant sloths, saber-toothed lions, and cave bears died out.

5. Globalization as a factor in the extinction of species

The proposed mechanism of mass extinctions is also quite acceptable for explaining the current situation with biodiversity on our planet. The processes of globalization, which are actively developing owing to man, largely act similarly to the influence of the unification of continents, reducing the impact of the geographic isolation factor. The activation of accidental (invasion) and deliberate (introduction) movement of animals and plants with the help of humans leads to competition between species that find themselves within the same ecological niche, with the inevitable subsequent extinction of less adapted ones. Thus, in recent decades, there has been a steady increase in the number of invasive species [24], which undoubtedly contributes to the extinction of the natives.

In the context of globalization, island populations are at particular risk. For example, the communities of the islands of Madagascar, New Zealand, and many other isolated territories, unique in terms of biodiversity, are constantly threatened by the introduction of new species.

First of all, representatives of animals are threatened. Plant species are less sensitive to competitive exclusion and more resistant to land consolidation because the spread of plant spores and seeds is global due to birds and atmospheric transport.

6. Discussion

Mass extinctions of biological species represent one of the obscure problems of evolution, perhaps overdramatized. Nevertheless, the available paleontological data on biodiversity "depressions" in some periods of the history of the biosphere require clarification of the causes of such events.

The proposed fairly universal hypothesis of mass extinctions is capable of explaining the mechanism of reductions in biodiversity at certain stages of the history of the biosphere without being too dramatic.

They are the tectonic movements of Earth's crust, which cause periodic consolidation and separation of lithospheric plates and, accordingly, change the degree of geographical isolation of species, which are proposed to consider as the major factor triggering the process of mass extinctions of species, which have repeatedly occurred in the history of the biosphere.

The mechanism of extinctions is implemented according to the competitive exclusion rule due to the growth of interspecific competition and the death of less adapted species in the conditions of a decrease in the diversity of ecological niches and the falling of different species into one ecological niche when lithospheric plates unite. This is especially clearly seen in the formation of supercontinents. With the separation of the lithospheric plates, a new progressive jump in speciation occurs, since the most competitive surviving species participate therein.

The removal of the isolation factor also explains the phenomenon of the Late Pleistocene extinction of megafauna in Northern Eurasia and North America as a result of the transgression and regression of the sea and the periodic unification of the old and new worlds.

It should be noted that data on the loss of plant species during periods of mass extinctions are usually less significant, which is associated with their lower migratory activity and, accordingly, with a significantly lower increase in competition when tectonic plates unite. For example, the periodic consolidation of the North American and Asian continents during the formation of the Bering Isthmus did not change the situation in the struggle for a "place under the Sun" for plants, and the exchange of their seeds and spores, for example, with the help of birds and wind, occurs even in the absence of the isthmus.

The tectonic hypothesis of the causes of mass extinctions also explains the long duration of the periods of mass extinctions, due to the fact that the extinction of species due to interspecific competition occurs during the change of a large number of generations.

Human activity, turning into an increasingly powerful geological factor, like the consolidation of lithospheric plates, has significantly reduced the isolation degree of biological species. During the period of globalization, due to introduction and invasion, migration processes have significantly accelerated, which inevitably leads to an increase in interspecific competition, to a decrease in biodiversity, and finally to a new mass extinction. There is no doubt that the process of global chemical and physical anthropogenic pollution of all components of the biosphere and the reduction of habitats of native species are superimposed on the said cause of, in fact, anthropogenic biological pollution.

The catastrophic natural processes considered in the literature as the primary source of mass extinctions undoubtedly played and continue to play a large role in reducing the number of species inhabiting the affected regions, but cannot fully explain the regularity of manifestations of global depressions in biodiversity dynamics and their duration.

Such cycles of mass extinctions of species during the unification of continents and jumps in speciation after the separation of the continents are evolutionary progress for the biosphere since the most surviving competitive species participate in a new round of speciation.

For final conclusions about the reality of the considered provisions, detailed comparisons of the periods of collision of tectonic plates (not only with the formation of supercontinents) and data on the biodiversity dynamics during these periods are required. At the same time, there are considerable limitations both from the side of dating the consolidations of tectonic plates and from the standpoint of assessing biodiversity during these periods.

Despite increasing scientific possibilities, many aspects of plate tectonics and the supercontinent cycle "cannot be demonstrated; even the number of supercontinents that may have existed prior to Pangea remains a matter of debate" ([16], p. 892). On the other hand, "any conclusions may be erroneous only because the paleontological record is a very small and nonrandom sample, which does not allow obtaining reliable information about the total ancient biota at a certain taxonomic level, and unreliable not only quantitatively, but even qualitatively" ([9], p. 8).

7. Conclusion

The considered material illustrates one more aspect of the influence of geological factors on the evolution of life on Earth. At the same time, the tectonic hypothesis of the causes of mass extinctions of species to the maximum extent, in comparison with other hypotheses, explains the features of these global phenomena, in particular:

1. the large proportion of extinct marine species, since when the plates merge, the area and diversity of shelf territories sharply reduce;

2. lack of reliable data on the extinction of plants, perhaps due to their less migratory ability, but greater globality due to the spread of their spores and seeds around the world by birds and atmospheric currents;

3. the jump in biodiversity following the extinction due to the increase in the diversity of ecological niches during the separation of the plates due to the participation of the most competitive species, which won;

4. long duration of mass extinctions due to the low speed of movement of lithospheric plates and the significant duration of the replacement of species in the process of interspecific competition;

5. recurrence of mass extinctions during the next association of lithospheric plates, which act as a trigger for increased competition, extinction of species, and subsequent growth of biodiversity when the plates are separated.

Further refinement of the periods of mass extinctions and periods of lithospheric plates joining divergence will make it possible to create a more accurate concept of the role of geological factors in the evolution of species and the dynamics of biodiversity.

Acknowledgments and funding sources

The work was supported by the state assignments of Lomonosov Moscow State University (AAAA-A16-116042010089-2) and the Institute of Basic Biological Problems RAS (FMRM-2022-0029).

Author details

Valeriy V. Snakin[1,2]

1 Lomonosov Moscow State University (Earth Science Museum), Moscow, Russian Federation

2 Institute of Basic Biological Problems of Russian Academy of Science, Moscow Region, Russian Federation

*Address all correspondence to: snakin@mail.ru

References

[1] Polishchuk LV. The three-quarter-power scaling of extinction risk in late Pleistocene mammals, and a new theory of the size selectivity of extinction. Evolutionary Ecology Research. 2010;**12**:1-22

[2] Burlatskaya SP. Archaeomagnetism. In: Structure and Evolution of the Earth Magnetic Field. Moscow: GEOS; 2007. p. 343 (in Russian)

[3] Nikolaev VI, Alekseev AO, Mishenko EN, Kuznetsova TV. Geochemical study of later Pleistocene Yakut mammoth tusks. Modern scientific research: Innovations and experience. Interdisciplinary Institute "Science and education". 2015;**5**(12):101-104 (in Russian)

[4] Greshko M, National Geographic staff. What Are Mass Extinctions, and What Causes them? IOP Publishing PhysicsWeb; 2019. Mass extinction facts and information from National Geographic [accessed on 29 June 2023]

[5] Scoville H. The 5 Major Mass Extinctions IOP Publishing PhysicsWeb. 2020. Available from: https://www.thoughtco.com/the-5-major-mass-extinctions-4018102. [accessed on 13 June 2022]

[6] Darvin C. The Origin of Species by Means of Natural Selection, or the Preservation of Favoured Races in the Struggle for Life. London: John Murray; 1859. p. 574

[7] Krassilov VA. Nature Conservation: Principles, Problems and Priorities. Moscow: VNIIPriroda; 1992. p. 173 (in Russian)

[8] Fedonkin MA. Biosphere: The fourth dimension. Priroda [The Nature]. 1991;**9**:10-18 (in Russian)

[9] Alekseev AS, Dmitriev VY, Ponomarenko AG. The Evolution of Taxonomic Diversity. Moscow: GEOS; 2001. p. 126 (in Russian)

[10] Corso JD, Gianolla P, Rigo M, et al. Multiple negative carbon-isotope excursions during the Carnian pluvial episode (late Triassic). Earth-Science Reviews. 2018;**185**:732-750. DOI: 10.1016/j.earscirev. 2018.07.004

[11] Ceballos G, Ehrlich PR, Barnosky AD, García A, Pringle RM, Palmer TM. Accelerated modern human–induced species losses: Entering the sixth mass extinction. Science Advances. 2015;**1**(5):1-5

[12] Racki G. A volcanic scenario for the Frasnian–Famennian major biotic crisis and other late Devonian global changes: More answers than questions? Global and Planetary Change. 2020;**189**:103-174. DOI: 10.1016/j.gloplacha.2020.103174

[13] Snakin VV. Mass extinctions in biosphere history: Another hypothesis. Izvestiya RAN. Ser. Geograph [Proceedings of the Russian Academy of Sciences]. 2016;**5**:82-90. DOI: 10.15356/0373-2444-2016-5-82-90 (in Russian)

[14] Snakin VV. Lithospheric plate tectonics and mass extinctions of biological species. IOP Conference Series: Earth and Environmental Science. 2021;**946**:1-7. DOI: 10.1088/1755-1315/946/1/012009

[15] Speculated Causes of the Permian Extinction. Hooper Virtual Paleontological Museum. Retrieved 16 July 2012 (Causes of the Permian Extinction (park.org) [accessed on 25 July 2022]

[16] Nance DR. The supercontinent cycle. In: Encyclopedia of Geology. 2nd ed. Publisher: Academic Press; 2021

[17] Domeier M, Van der Voo R, Torsvik TH. Paleomagnetism and Pangea: The road to reconciliation. Tectonophysics. 2012;**514-517**:14-43

[18] Nance DR, Murphy BJ. Origins of the supercontinent cycle. Geoscience Frontiers. 2013;**4**:439-448

[19] Meert JG. What's in a name? The Columbia (Paleopangaea/Nuna) supercontinent. Gondwana Research. 2012;**21**:987-993. DOI: 10.1016/j.gr.2011.12.002

[20] Nance RD, Murphy JB, Santosh M. The supercontinent cycle: A retrospective essay. Gondwana Research. 2014;**25**:4-29. DOI: 10.1016/j.gr.2012.12.026

[21] Scotese ChR. Paleomap project. 2014. Available from: Future +250 (scotese.com) [accessed on 13 June 2022]

[22] Snakin VV. Ecology, Global Environmental Processes and Biosphere Evolution: Encyclopedic Dictionary. Moscow: Moscow University Press; 2020. p. 528. Available from: https://www.rfbr.ru/rffi/ru/books/o_2101835 [accessed on 13 June 2022] (in Russian)

[23] Lorenzen ED, Nogués-Bravo D, Orlando L, et al. Species-specific responses of late quaternary megafauna to climate and humans. Nature. 2011;**17**(479):359-364

[24] Global Biodiversity Outlook 3. Montreal: Secretariat of the convention on biological diversity. 2010; 95

Chapter 6

Perspective Chapter: Ostracoda

Hanan Zwair

Abstract

Ostracoda are small, bivalved crustaceans that have thrived on Earth for 500 million years. They inhabit marine and freshwater environments, and their well-preserved fossil record offers insights into past ecosystems and paleoenvironmental conditions. Ostracod shells, known as valves, are made of calcium carbonate or chitinous material, enclosing the body in two separate halves. The valves display diverse shapes and ornamentation, reflecting the wide variety of ostracod species. Internally, ostracods possess a developed musculature, antennae for sensing the environment and capturing food, mandibles for chewing, and maxillae for filtering food. Their body cavity houses the digestive, circulatory, and reproductive systems. Ostracod fossils are found in marine, lacustrine, and terrestrial deposits, providing opportunities to study their morphology, distribution, and evolution. Their rapid evolutionary rates and broad geographic distribution make ostracods valuable for biostratigraphy. By analyzing ostracod assemblages in fossil sediments, researchers can reconstruct ancient ecosystems and track environmental changes. Ostracods serve as indicators of environmental conditions such as water temperature, salinity, and oxygen levels. Studying ostracod fossils informs our understanding of past environments and climate change.

Keywords: ostracoda, morphology, ecology, fossils, biostratigraphy

1. Introduction

Freshwater and brackish water Ostracoda are small-sized crustaceans with a wide and large spread in all aquatic environments: streams, rivers, ponds, swamps, and shallow and deep ones. Despite that, it has received little attention from specialists because of its small size and the difficulty of dealing with the carapace [1]. Its importance, theoretically and practically, has not been evaluated so far, despite its role in the food chain. Ostracoda has enormous fossil records, which are important in geological studies [2].

The presence of Ostracoda is related to environmental factors and therefore related to the physical and chemical properties of the aquatic environment, so chemical and physical analyzes can determine the characteristics of these habitats [3]. The precise characteristics of the body and its appendages for classification are considered an impediment to environmental studies, as it is impossible to study any organism without classifying it to the species [4].

Ostracoda is small crustaceans mostly in length (5–0.3 mm), and the carapace completely covers their body. It is found mainly among plants and their remains, and appears at first sight as a small, kidney-shaped seed and can be easily distinguished by its carapace consisting of two shutters, which are closed by adductor muscles. It includes appendages inside, and the body does not contain any divisions [5–8]. The carapace surrounding it gives it a high specificity, as it resembles in its shape the bivalve mollusks; a number of them are plankton forms, but most of them live on the bottom where they are crawling or swimming near the surface of the water. Most of the carapaces carry their eggs in the brood pouch, which is the space between the dorsal carapace and the body.

Ostracoda is found all over the world and in both marine and non-marine environments and abound in all types of freshwater environments, whether permanent or temporary stagnant or running. It is considered one of the largest and most diverse groups of crustaceans and currently includes about 30,000–50,000 living species and 30,000 fossilized species [9, 10].

As for the fossilized ones, they are found near all geological layers. They predominated in the Silurian period and were found at a depth of 9 cm [11].

Many Ostracoda species are environmentally specialized and considered indicators of various environmental conditions, mainly temperature, depth, and salinity [12]. It is also considered an important part of the aquatic ecosystem and a unique part of the huge diversity of crustaceans. The small size of its crust and the difficulty of classifying and dealing with its crust reduce its popularity among naturalists [13].

The Limnocytheridae and Cypridacea families are the most common Ostracoda families in non-marine environments and are often considered to be of great interest and value in the knowledge of palaeoenvironment [14, 15].

Drakes prefer to live in clean, oxygenated water over polluted water, are more active in the evening, and do not prefer high light [16].

2. Classification of ostracoda

Phylum: Arthropoda, Subphylum: Crustacea, Class: Ostracoda, The issue of categorizing Ostracoda into Orders, Suborders, Superfamilies, and Families still needs more discussion among specialists.

Ostracoda is divided into four main orders:

Order: Paleocopida Henning, 1953. Fossilized species only.

Order: Myodocopida Sars, 1866. Marine species only.

Order: Platycopida Sars, 1866. Marine species and a few of them are marine and fresh species.

Order: Podocopida Sars, 1866. Brackish.

Podocopida and all types of freshwater dromedaries fall under the order Podocopida and are divided into three main families:

(Darwinulidae, Cyprididae, and Cytheridedae)

The three families can be distinguished by:

1. The similarity or lack of similarity of the three pairs of thoracopoda.

2. The number and order of muscle scars for each family.

3. The exopod of the second Antenna is modified into a scaly structure that carries a number of cilia or a long spiny structure (**Figure 1**) [17].

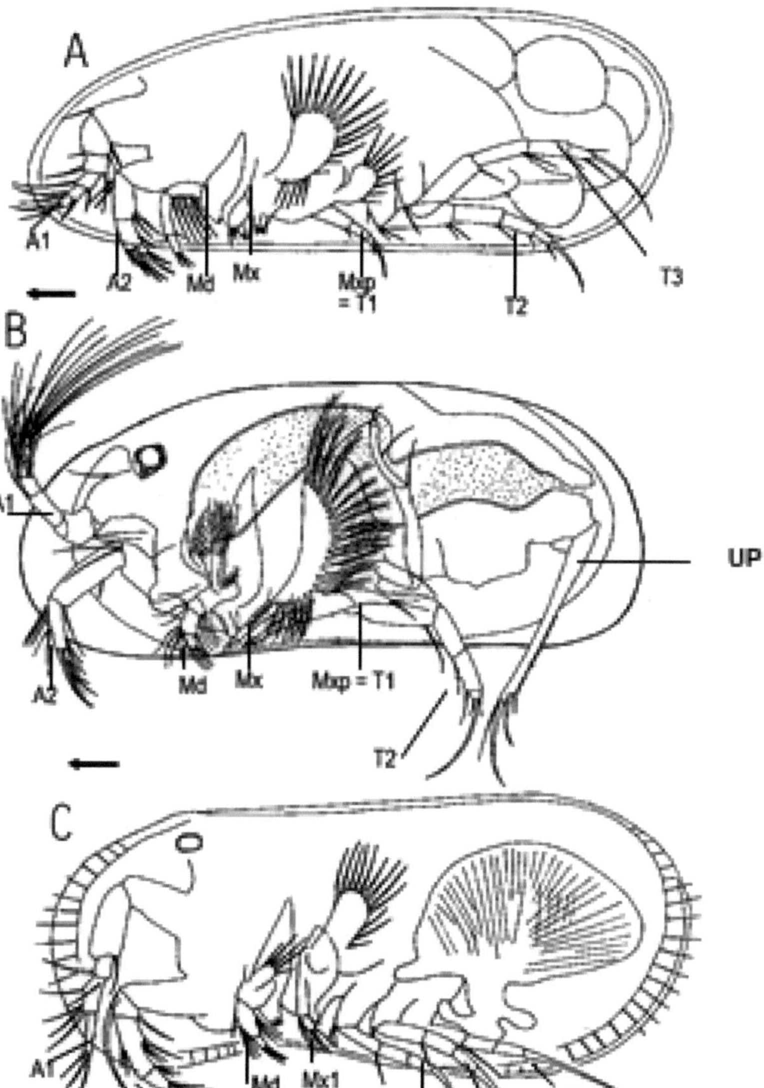

Figure 1.
Three Superfamilies: A—Darwinula strevensoni (Darwinuloidea), B—Heterocypric reptans (Cypridoidea), C—Limnocythere sanctipatricii (Cytheroidea). A&B. female, C. male [17].

3. General structure of ostracoda

3.1 Carapace

The carapace, consisting of two valves, covers the body of the Ostracoda completely, and the body consists of two regions: the head or cephalon and the thorax, separated by narrow partitions [18].

The two loose ends of the shutter snap together or overlap, creating a small loose end. A thick bundle of strong adductor muscles passes through the center of the body,

Figure 2.
Pattern of adductor muscle scar order: Podocopoda [20].

which contracts to close the two valves. The complete closure of the carapace provides the animal with a strong defense system against many types of predators and temporary drought conditions that occur in the habitat [19].

The two valves close by the adductor muscles (**Figure 2**) pass through the central part of the body and are attached to the inner surface of the calcified outer plates, where characteristic scars are formed. The adductor scar pattern is another important taxonomic feature in addition to the adductor scar pattern [20].

The valve consists of two double-folded excrescence of the body wall, which is two outer and inner lamellas, and there are in the space or space separating the two plates, perhaps reproductive organs or the alimentary canal, according to the orders and families [21].

The outer surfaces of the valves may be smooth, embossed with small pits, pitted, ornamented with reticulate ridges or spins [20]. The valves cover channels with radial or normal holes. The importance of the nature of the channels is in dividing species groups and as an idea for identifying and isolating species [22].

3.2 Appendages

The carapace appendages are specialized for locomotion, feeding, and reproduction. Appendages differ greatly in form and function, without a doubt, with different habitats, environments, and lifestyles. The number of Podomeres and Setae that make up the appendages is very important in the classification.

3.3 First antenna

Its function is sensory and kinetic, and it is uniramous and consists of (5–8) segments. The exopod is absent or reduced to a single inconspicuous palp. In swimming species, the palp is long, flexible, and thin, containing a number of natatory setae. As for crawling and climbing ones, they have short claw-like setae.

3.4 Second antenna

It is considered the main kinetic structure of walking and has great taxonomic importance in the Podocopida; the inner Endopod is well developed and consists of (3–4) articulated segments bearing a variety or group of setae with claw ends. As for the Exopod, it is reduced to a scale structure that usually bears three setae. The first segment of the Endopod of most of the family Cypridoidea bears (1–5) sensory setae.

3.5 Mandible

It contains a masticatory process with strong teeth and a textured attachment that usually bears a well-developed vibratory plate. Two digestive parts are attached to the lower lip and located in the oral cavity.

3.6 Maxilla

It consists of palp having two segments and three masticatory processes usually bearing a vibrating plate, which generates a continuous stream of water through the body for respiration. **Maxilla** are located directly behind **Mandible** and have two functions, respiratory and feeding.

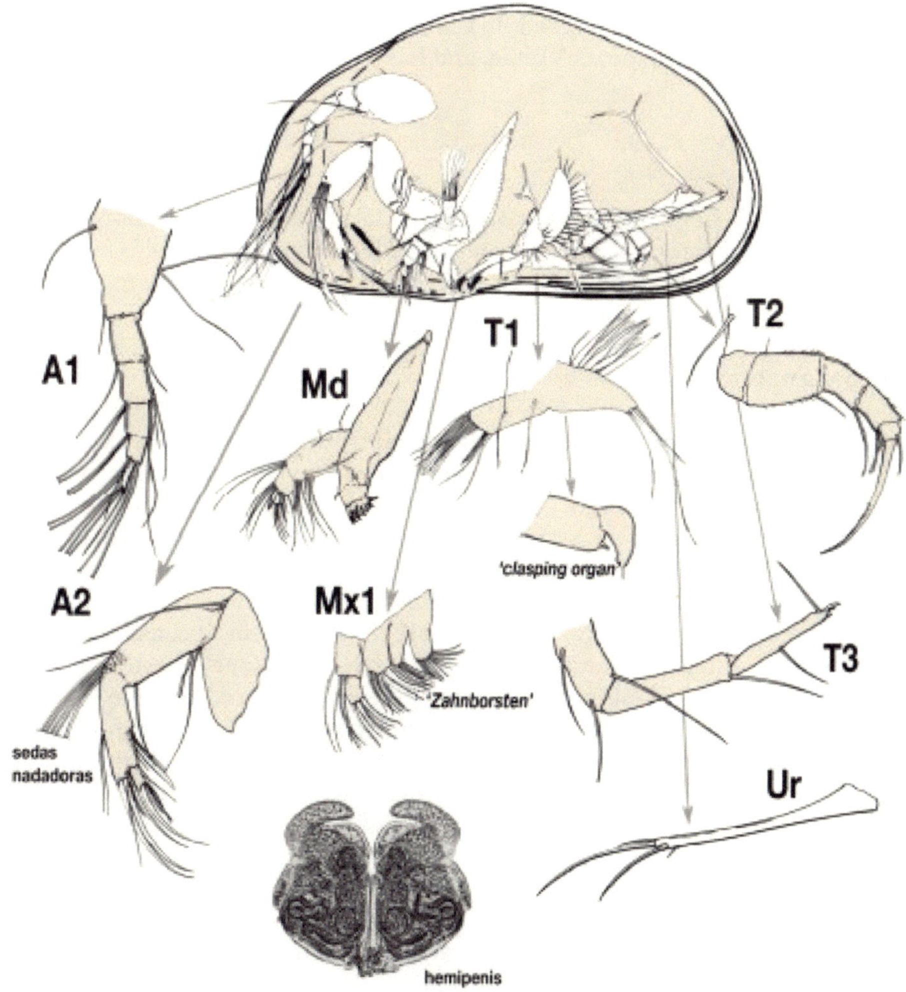

Figure 3.
Appendages of Podocopida [23].

3.7 First Thoracopod

It is a distinct structure in its form and function, as it works either as walking legs as in (Cytheroidea) or modified into a feeding organ as in (Darwinuloidea and Cypridoidea) and in males of the genus (Cypridoids) the Endopod modified into a clasping organ.

3.8 Second Thoracopod

It consists of all members of the Ostracoda of 4–5 segments, which usually bear a developing apical claw. It acts as a walking man.

3.9 Third Thoracopod

In superfamilies, Cytheroidea and Darwinuloidea, they are in the form of walking legs and resemble the **second Thoracopod.** In Cypridoidea, they are modified into cleaning legs that are curved or arched. In Candonidae, the distal end of the cleaning legs is slightly modified and bears 3 setae, and has taxonomic importance at the level of genera.

3.10 Uropod

Rod-shaped, flat-sided, undivided, bearing distal claws and two or more setae. In the family Cypridopsinae, it is partially reduced (flagelliform) in females, while males are completely reduced (**Figure 3**) [1, 17, 18].

4. Internal anatomy

Podocopida lack gills or any specialized respiratory organ, and gas exchange occurs by entering the body surface and the inner lamella, where gases are exchanged through the chitin layer of the inner surface of the valve [24]. Continuous ventilation of the respiratory plates equips the animal with oxygenated water and displaces carbon dioxide (CO_2) and the rest of the materials thrown through the valves. It also lacks the heart, as blood circulation occurs through muscle contraction in the body wall or the blood vessels themselves. As for the digestive system, it consists of the oral cavity, esophagus, anterior intestine, which can act as the stomach, posterior intestine, anus, which is located on the dorsal side of the uropod. In addition to the presence of the hepatopancreas, it collects the secretions of a pair of digestive glands and is located between the two plates of the valves. The alimentary canal or intestines may sometimes be visible through the carapace. The central nervous system comprises the cerebrum, the circum-esophageal collar of fused ganglia, and the ventral chain of ganglia. The brain supplies the nerves to each of the eyes, the lower layer of the valve, and the pairs of the first and second tentacles, and the rest of the appendages are supplied by the ventral chain of the ganglia. The eyes are above the base of the first Antenna and can be distinguished as a dark spot, usually in living animals. The eye includes three optical cups, which are usually fused. Ostracoda has two different visual systems, a single median eye and two pairs of compound eyes [4].

5. Reproductive system

The male reproductive system is distinguished in its complexity and size. It represents about a third or a quarter of the size of the body. It includes a pair of testes. Each testis consists of four long tubular branches between each valve's membrane, which unite to form the vas deferens, the seminal vesicle, and Zenker organ, which acts as a pump to push the sperm into the penis forcefully. In the female, the reproductive system occupies half the size of the body. It is simple in structure and includes a pair of tubular ovaries paired, oviduct, seminal receptacle, seminal receptacles, and genital opening.

5.1 Modes of reproduction

There are four modes of reproduction for Ostracoda groups depending on the presence or absence of males:

1. The first group: includes some species of both sexes, such as *Candona and Cypricercus*. Males are absent; the reproduction pattern is parthenogenesis.

2. The second group includes the genera *Cyprinotus, Candona*, *Darwinula, Ilyocypris*; males are rare, and the reproduction pattern here is also parthenogenesis.

3. The third group: includes the genera *Potamocypris, Darwinula, Cypridopsis, Cypricercus, Darwinula*, the presence of males here in small numbers, and the mode of reproduction here is sexual and pathogenic

4. The fourth group: includes the species of genera

5. *Cyclocypris, Cypria Candona Cypricercus. Cyprois, Limnocythere, Physocypria, Notodromas,* males are always present, and the reproductive pattern is sexual.

The eggs of the Ostracoda are spherical with a thin bilateral wall; their colors range from yellow, white, red, orange, and green. The eggs are transported by fish, birds, wind, or any other way [1, 17]. Podocopida lack the stage of the larval or coastal stage pelagic larval, so their migration or movement is determined by active motile elements and by the nature of the prevailing benthic current [25–28].

In environmental experiments, it was shown that the eggs of the Ostracoda retain their vitality even after being exposed to dry conditions at 20 and 40°C and after being exposed to freezing conditions at 18°C [28].

And the eggs of freshwater Ostracoda are highly resistant to the damage they are exposed to and can withstand harsh conditions [29].

And to reach the adult stage, the Ostracoda, like the rest of the crustaceans, suffer from molting several times (8 times) between their growth stages; immediately after that, the two valves harden in the form of new layers by taking compounds from the surrounding water [30].

Before molting occurs, the new epicuticle begins to harden, and during this process, trace elements merge and unite in the crust; and this phase is somewhat rapid, and these chemical elements, from which the crust is formed, can diagnose some phenomena in marine and fresh environments since the time the new crust was formed, such as pollution [31].

6. Feeding

The food of the Ostracoda consists mainly of bacteria, molds, algae, and fine detritus, and darts have several ways of feeding; most of them are filter feeding, as the beatings or movements generated by the cilia of mandibular palps, maxillary processes, and vibratory plate for the first thoracopod, a continuous stream of water is generated between the two valves, which enables it to catch the small particles with their Bristel appendages and deliver them to the mouth [1].

And some species of carnivores are predatory, Detritivorous herbitory, predaceous carnivore, or scavengers [32].

And species of them are parasitic on other crustaceans, polychaete worms, echinoderms, or sharks [18].

7. Ecology of ostracoda

In the eighteenth century, studies were largely focused on taxonomy, but the environmental sensitivity of Ostracoda was unknown among most environmentalists [33].

Ostracoda that live in continental waters are sensitive to various aquatic environments. Some species are found in springs, rivers, streams, ponds, or lakes, and others can be found in more or all environments and habitats, whether at the surface or depths, and they are endemic or alkaline [16, 34–36].

The presence of Ostracoda can be controlled by constant factors such as food availability, depth, pH, dissolved ions, oxygen concentration, temperature, concentrations of dissolved substances, salinity, electrical conductivity, ionic strength, and calcic saturation index [3, 7, 36–39].

The small size of the Ostracoda, its large numbers, and its sensitivity to environmental conditions make it ideal for determining the disturbance and change in lake environments since the non-marine dracaena is often used as an indicator of seasonal changes that occur in large lakes, as well as the dracaena is a source of information about past seasonal changes in the chemical and water temperature [40, 41].

The non-marine groups that live in freshwater are controlled by calcium ions Ca2+, calcium carbonate Hco3, and magnesium, Mg2+, to a typical extent, and cannot live in water dominated by sodium ions, Na+, chlorine, or other types of ions. These main ions play an important role in knowing the sensitivity of Ostracoda to dissolved compounds [42, 43].

The calcite crust of the Ostracoda is a source of data for stable and radiogenic elements and trace metals. These data test and refine the structure and organization of Ostracode paleohydrological [16, 44].

Species respond to temperature and other environmental changes, such as oxygen concentration and pH (pH, O_2), as they affect their succession and survival. When food is available, growth rates increase with increasing temperature. On the contrary, food rates that support growth at low temperatures may be insufficient or adequate at high temperatures [37, 41, 45].

The different species of Ostracoda require a difference in the environmental conditions, such as the change in temperature and oxygen, for the molting process to occur for growth. The temperature is very important for the stages of the life cycle of the Ostracoda (e.g., hatching, growth, and reproduction) [33, 46].

Salinity, ionic compounds, and water contents are important environmental factors and are considered the main dominant that affect the presence and distribution of marine and non-marine (brackish and fresh). They showed the importance of salinity in the life cycle of organisms such as Ostracoda. The external shape of the valve in the Ostracoda is in terms of thickness or decoration [47, 48].

Some species of Ostracoda may not be affected by the surrounding environmental conditions such as high salinity [49].

Oxygen is considered a catalyst that works for the survival of Ostracoda like other living organisms; some of them have no resistance to withstand the sudden lack of oxygen, most of the freshwater Cypridacea is abundant in stagnant water with little oxygen, and some of it is formed during harsh periods and long hot weather [50].

8. Fossil Ostracoda and paleoecology

Ostracoda that has a calcite valve easily turns into fossils in the sediments of lakes or wet places or any place where the animal resides or lives [21, 51].

The oldest rocks found in which ostracoda fossils date back to the Upper Cambrian period belong to the family Leperditidae of the order Leperditicopida [52].

Ostracoda considers at paleontologists, stratigraphies, and environmental indices important applications in oil and gas exploration and makes them an excellent subject in evolutionary studies [18].

In freshwater environments, ostracods are commonly found in lake and pond sediments and can provide insights into the changes in water chemistry, temperature, and sedimentation over time. The remains of ostracods, together with other microfossils, can help reconstruct past environments and climates, providing information on long-term environmental and biotic changes [53].

Important of Fossil ostracods are not only for their value in reconstructing past environmental conditions, but also for determining the relative ages of sedimentary rocks. Because ostracods have evolved so rapidly and their remains are abundant, they make excellent tools for biostratigraphy, allowing geologists to correlate rocks over large areas and between widely separated regions [54].

8.1 Stratigraphic significance and distribution

The morphological characteristics of fossil ostracods have significant implications for understanding the stratigraphic distribution of these creatures. Because ostracods have evolved rapidly and their morphology is sensitive to changes in environmental conditions, they make excellent tools for biostratigraphy, allowing geologists to correlate sedimentary rocks over large areas and between widely separated regions [55]. In addition, the variations in ostracod morphology over time can be used to trace the evolutionary history of these creatures and to infer patterns of diversification and extinction. For example, the rapid radiation of ostracods during the Paleozoic era can be linked to the onset of ecological and environmental changes during that time, including the evolution of new habitats and the rise of new predator-prey relationships. Fossil ostracoda have a widespread stratigraphic distribution, and their remains have been used as biostratigraphic markers to correlate sedimentary successions over broad spatial ranges. The key factors that influence the stratigraphic distribution of fossil ostracoda are their evolutionary history, ecological preferences, and environmental conditions. Ostracods have evolved rapidly, with many new species originating during the periods of ecological and environmental change. For example, the rapid radiation of ostracods during the Paleozoic era can be linked to the onset of ecological and environmental changes during that time, including the evolution of new habitats and the rise of new predator-prey relationships [56].

Ecological preferences of ostracod species can also affect their stratigraphic distribution. Ostracods that prefer freshwater environments are generally more abundant in freshwater sediments, while those that favor marine or deep-sea environments are more prevalent in marine sediments [54]. Environmental factors such as temperature, salinity, and pH can similarly shape ostracod distribution patterns. For instance, the presence of different ostracod species in marine environments can be used to infer salinity and temperature values of past seawater. The Mesozoic era is also famous for its diverse ostracod fauna, with many notable groups appearing during this time. For example, the Lower Jurassic period saw the appearance of the parasitic cypridinids, which were found in deep-sea sediments, while the Cretaceous period witnessed the rise of the suborder Podocopina, which includes many freshwater ostracods [57].

8.2 Global distribution of ostracodes

Ostracodes are aquatic and semiaquatic organisms, and their distribution varies widely based on ecological factors such as temperature, light levels, and water chemistry. These tiny crustaceans can be found in almost every aquatic environment, including lakes, streams, rivers, and oceans, and they are a key component of many food webs.

The distribution of ostracodes varies based on age and living stages. For example, ostracodes in their early developmental stages can be found in shallow areas with higher oxygen levels, while adults tend to inhabit deeper waters or even the benthic zone. Additionally, ostracodes are known to be sensitive to changes in water temperature and salinity, meaning their distribution can also be influenced by seasonal shifts or climate change.

In terms of geographic distribution, ostracodes are found in freshwater and marine environments all over the world, from the tropics to the poles. Some species are more abundant in certain regions, such as the Arctic and Antarctic, where they have adapted to extreme conditions. Researchers have also documented specific ostracode species associated with particular geological formations, which has implications for geochronology and stratigraphic studies [58, 59] (**Figure 4**).

8.3 Stratigraphic correlation

The stratigraphic correlation of fossil ostracoda is a crucial aspect of paleontological research, as it allows geologists to establish relative ages and correlations of sedimentary rocks over wide spatial areas. In this chapter, we explore the techniques used for stratigraphic correlation of fossil ostracoda, their applications, and their

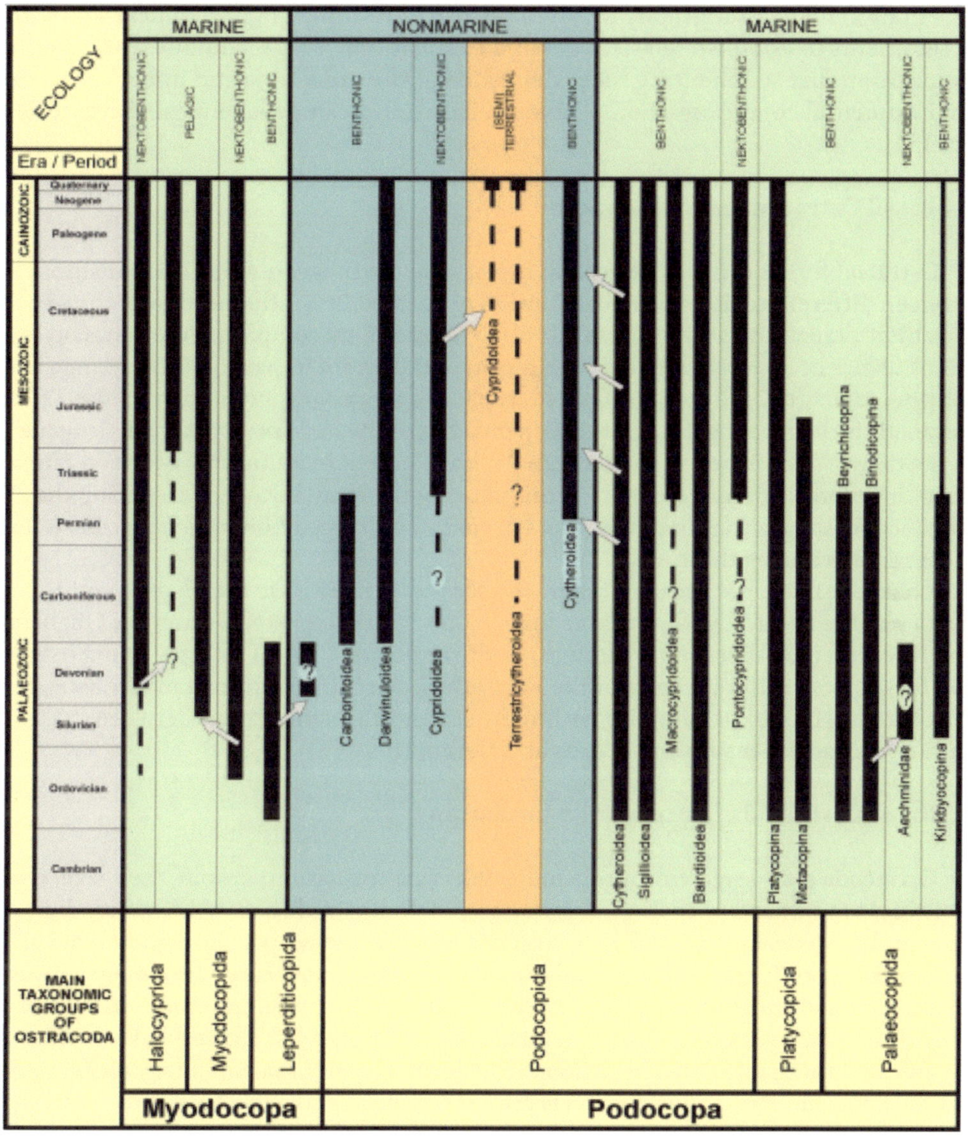

Figure 4.
Stratigraphical timelines of the major ecological radiations of the Ostracoda [60].

limitations. The primary techniques used for stratigraphic correlation of fossil ostracoda include biostratigraphy, lithostratigraphy, chronostratigraphy, and chemostratigraphy. Biostratigraphy is the most commonly used technique and relies on the identification and correlation of fossil assemblages. Lithostratigraphy involves the correlation of rocks based on their lithological characteristics, such as grain size and mineralogy. This technique is useful when fossils are absent or when other correlation methods are not applicable. Chronostratigraphy is the application of standardized geological time scales to establish the age and correlation of sedimentary rocks. This technique relies on the identification of stratigraphic boundaries and the allocation of rock units to different intervals of geological time.

Finally, chemostratigraphy involves the use of isotopic or geochemical signatures in sediments to establish relative ages and correlations. This technique works on the assumption that sedimentary rocks deposited at the same time and under the same environmental conditions should have similar isotopic or geochemical characteristics [54].

8.4 Fossil Ostracoda in biogeography

Ostracod fossils can also be used to investigate the biogeographic relationships between different regions. The distribution of ostracods is influenced by factors such as habitat availability and dispersal ability. Therefore, by comparing the ostracod faunas of different regions, scientists can test hypotheses about past biotic exchange and continental drift. **Case studies**: One example of a biogeographic study that used fossil ostracods is the investigation of the dispersal of freshwater taxa from the Paleogene of northern South America to southern Africa. This study found that ostracod faunas from these two regions were more similar during the early Eocene than during the late Eocene, suggesting that there was a period of enhanced biotic exchange between these regions during this time.

Another example is the use of ostracod fossils to investigate the biogeographic history of the Paratethys Sea, a large but shallow sea that existed during the Oligocene and Miocene (~30–5 million years ago) in the region of Central Europe. Using ostracod fossils from different parts of the Paratethys, scientists were able to reconstruct the changing biotic connectivity within the sea and the connections between the Paratethys and other seas and oceans at different times [61–64].

8.5 Role of ostracoda in earth crust formation

Ostracoda play a significant role in Earth's crust formation, despite their size, they contribute to important ecological processes, including sedimentation, mineralization, and biogeochemical cycling. Ostracods have a direct impact on sedimentation processes by producing and accumulating their shells in aquatic environments. These shells, composed mainly of calcium carbonate or silica, contribute to the formation of sediment layers over time. The accumulation of ostracod shells, along with other organic and inorganic particles, leads to the development of sedimentary rocks, such as limestone and shale. This process is crucial for the overall structure and composition of Earth's crust [18].

Ostracoda and Biomineralization: Ostracods have the unique ability to biomineralize their shells, a process that involves the deposition and organization of minerals within their exoskeletons. This biomineralization process contributes to the formation of hard structures that can withstand physical and chemical weathering, eventually becoming part of Earth's crust. The controlled precipitation of minerals within ostracod shells aids in stabilizing sediments and contributing to the development of calcareous and siliceous formations.

Ostracoda and biogeochemical cycling: Ostracods play an essential role in biogeochemical cycling by participating in nutrient cycling and remineralization processes. They contribute to the breakdown of organic matter and the recycling of nutrients in aquatic ecosystems. The activities of ostracods influence the availability and distribution of essential elements, such as carbon, nitrogen, and phosphorus, in the environment. These processes, in turn, impact the overall nutrient balance and the formation of sediments in Earth's crust [56].

9. Conclusions

1. The morphology of ostracoda, characterized by their bivalved shells and complex internal structures, is highly diverse and well-preserved in the fossil record. The varied shapes and ornamentation of their valves reflect the wide range of ostracod species. The detailed study of their morphology provides insights into their evolutionary history and adaptations to different environments.

2. Ostracods inhabit both marine and freshwater ecosystems, and their fossils provide valuable information about past environments and ecological conditions. Ostracods are highly sensitive to environmental parameters such as temperature, salinity, and oxygen levels. By analyzing ostracod assemblages in fossil sediments, researchers can reconstruct ancient ecosystems and track changes in environmental factors over time.

3. Fossils: Ostracod fossils are abundant and widely distributed in sedimentary deposits. Their excellent preservation in the fossil record allows for detailed examination of their morphology, providing valuable data for paleontologists. Ostracods have a rapid evolutionary rate and a wide geographic distribution, making them useful for biostratigraphy and dating sedimentary layers based on their fossil assemblages.

4. Paleoenvironmental Reconstruction: Ostracod fossils serve as important indicators of past environmental conditions. Changes in ostracod species composition and diversity can reveal fluctuations in water temperature, salinity, and other environmental parameters. By studying ostracod fossils, researchers can reconstruct past ecosystems, track climate change, and gain insights into the interactions between organisms and their environments throughout geological time.

5. Importance of Ostracod Fossils: Ostracod fossils play a significant role in paleontological research, contributing to our understanding of Earth's history and the evolution of aquatic ecosystems. Their widespread distribution, diverse morphology, and sensitivity to environmental changes make them valuable tools for paleoenvironmental reconstruction, biostratigraphy, and paleoecology studies.

Author details

Hanan Zwair
Department of Biology, College of Education for Pure Science, University of Karbala, Karbala, Iraq

*Address all correspondence to: hanan.mikhlif@uokerbala.edu.iq

References

[1] Pennak RW. Fresh-water invertebrates of the United States. Journal of Aquatic Ecosystem Stress and Recovery (Formerly Journal of Aquatic Ecosystem Health). 1955;7(1-2):126

[2] Bronshtein, Z. S. Fresh – Water Ostracoda. Translate by Nayar I. New Delhi: Oxonian Press Pvt. Ltd. 1947. p. 470

[3] Forester RM. Ostracode assemblages from in the western United States; implications for paleohydrology. The Memoirs of the Entomological Society of Canada. 1991;**155**:181-201

[4] Henderson PA. Freshwater Ostracodes. Pisces Conservation Ltd; 2002. p. 258

[5] Cole GA. Textbook of Limnology. 3rd ed. London: The C. V. Mosby, Company; 1983. p. 401

[6] Raymont JEG. Plankton and Productivity in the Oceans (2nd ed). 1980. p. 489

[7] Cronin TM, Holtz TR, Brouwers EMW, Briggs M, Whatley RC, Wood A. Arctic Ostracode database. Oceanographic Literature Review. 1996;**1**(43):92

[8] Huys R. An updated classification of the recent crustacea. Journal of Crustacean Biology. 2003;**23**(2):495-497

[9] Martens K. The Non-Marine Ostracods (Ostracoda) of Southern Africa. Freshwater Biology. Belgium: Royal Belgian Institute of Natural Sciences, vautierstraat 29; 2000

[10] Smith AJ, Park LE. An introduction to Ostracoda. In: Park LE, Smith AJ, editors. Bridging the Gap Trends in the Ostracode Biological Ad Geological Sciences. The Paleontological Society Papers. Vol. 9. 2003. pp. 9-10

[11] McConnaughey BH, Zottoli R. Introduction to Marine Biology. 4th ed. The C. V. Mosby Company London; 1983. p. 638

[12] Macan TT, Worthington EB. Life in Lakes & Rivers. Great Britain: William Collins Sons & Co. Ltd.; 1972. p. 320

[13] Edmondson WT. Freshwater Biology. 2nd. ed. Newyork: John Wiley & Sons, Inc.; 1959. p. 124

[14] Tatman S, Whately R. A palaeontological reconstruction of the ostracod member of the Blairmore formation, Alberta, Canada. In: Proc. 2nd. Europ. Ostracod. Meet. Univ. Glasgow. Scotl & . 23 Micrplaeontological Society 16th December, 1993. British Micropalaeontological Society London; 1996. pp. 35-38

[15] Whatley RC, Cusminsky GC. In: Riha R, editor. Quaternary Lacustrine Ostracoda from Northern Patagonia, Argentina. Ostracoda & Biostratigraphy; 1995. pp. 303-310

[16] Last WM, Tellev JT, Forester RM. Paleohydrology & paleochemistry of Lake Manitoba, Canada: The isotope & Ostracode records. Journal of Paleolimnology. 1994;**12**:269-282

[17] Meisch C. Freshwater Ostracoda of Western & Central Europe. Berlin: Spektrum Akademisher Verlog GmbH. Heidelberg; 2000. p. 513

[18] Athersuch J, Horne DJ, Whittaker JE. Introduction to ostracoda. In:

Brill EJ, editor. Marine & Brakisn Water Ostracods. Synopsis of the British Fauna. Vol. 43. 1989. p. 343

[19] Benson RH. The role of ornamentation in the design & function of the ostracode carapace. Geoscience & Man. 1974;**6**:417-457

[20] Horne DJ, Cohen A, Martens K. Taxonomy, morphology and biology of quaternary and living Ostracoda. The Ostracoda: Applications in Quaternary Research. 2002;**131**:5-36

[21] Benson RH. Biomechanical stability & sudden change in the evolution of the deep-sea ostracode pedeidonamicus. Paleobiology. 1983;**9**(4):398-413

[22] Coles GP, Whatly RC, Moguilevsky A. The ostracod genus Krithe form the Tertiary & Quaternary of the North Atlantic. Palaeontology. 1994;**37**(1):71-120

[23] Baltanás Á, Joanes MF. Order Podocopida. Revista IDE@-SEA. 2015;**74**:1-10

[24] Whatley R, Lethiers F. The use of Ostracoda to reconstruct the oxygen levels of late paleozoic oceans. Morine Micropaleontology. 1994;**24**:57-69

[25] Whatley R, Toy N, Moguilevsky A, Coxill D. Ostracoda from the south West Atlantic part I. the Falkland Islands. Revista Espanola De Micropaleontologia. 1995;**27**(1)

[26] Witte LJ. Taxonomy and Origin of Modern West African Shallow Marine Ostracoda. Amsterdam: Academische Pers; 1993. p. 201

[27] Whatley R, Boomer I. Upper oligiocene to pleistocone ostrocoda from Guyotss in the Western Pacific: Itoles & 71A, 872C, & 783B. In: Haggerty JA, Premolisliva I, Rack F, Mc Nutt MK, editors. Proceedings of the Ocean Drilling Program, Scientific Results. Vol. 144. 1995. pp. 87-96

[28] Angell RW, Hancock JW. Response of eggs of Heterocyparis in congruens (Ostracoda) to experimental stress. Journal of Crustacean Biology. 1989;**9**(3):381-386

[29] Szczechura J. Couses for speciation in ostracodes. N.J b.Geol.Palaönt.Mh. 1980;**7**:439-441

[30] Benson RH. Form, function, & architecture ostracode shells. Annual Review of Earth and Planetary Sciences. 1981;**9**:59-80

[31] Reyment RA. Case study of the statistical analysis of chemical composition exemplified by ostracod shells. Environment. 1996;**7**:39-47

[32] Balient SC, Whatley R. Marine ostracods as a chronoecological tool. Their application to the Andean Jurassic sequences of Argentina. XIII Congreso Geologico Argentineoy III congreso de Exploracion de Hidrocarbros, Actas; 1996;**5**:143-149

[33] Forester RM, Delorme LD, Ager TA. A Lacustrine Record of Late Holocene Climate Change from South-Central Alaska. Aspects of Climate Variability in the Pacific and the Western Americas. Vol. 55. 1989. pp. 33-40

[34] Lowry JK. Crustacea, the Higher Taxa. 1999. Available from: http://Crastacea.net/index.htm

[35] Nomiotko T, Szczechura J, Namiotko L. Ostracoda of the Eemian Interglucial At kruklanki in Ne Poland. Studia Quaternaria. 2003;**20**:3-24

[36] Smith AJ, Delorme LD, Forester RM. A lake's solute history from ostracodes:

Comparison of methods. In: Kharaka YK, Maest AS, editors. Water. Rock Interaction Proceedings of the 7th International Symposium on Water-Rock Interaction. Vol. 2. 1992. pp. 677-680

[37] Kontrovitz M. Ostracode shells as indicators of thermal history. In: Transactions Gulfcoast Assoc. Geological Societies, San Antonio, Texas. Vol. 37. 1987. pp. 383-391

[38] Dean WE, Forester RM, Bradbury JP. Early Holocene change in atmospheric circulation in the northern Great Plains an upstream view of the 8.2 ka cold event. Quaternary Science Reviews. 2002;**21**:1763-1775

[39] Szczechura J. Ecological control of ostracode distribution in the middle Miocene of paratethyes. In: Keen MC, editor. Proc, 2nd Europ. Ostracod. Meet. Univ. Glasgow, Scotland 23 Rd–27th July, 1993. London: Brit. Micropal. Soc.; 1996. pp. 99-106

[40] Park LE, Ricketts RD. Evolutionary history of the Ostracoda & the origin of Nonmorine faunas. In: Park LE, Smith AJ, editors. Bridging the Gap Trens in the Ostracode Biological & Geological Sciences. The Paleontological Society Papers. Vol. 9. 2003. pp. 11-36

[41] Smith AJ, Davis JW, Palmer DF, Forester RM, Curry BB. Ostracoda as hydrologic indicators in springs, streams and wetlands: A tool for environmental and paleoenvironmental assessment. In: Park LE, Smith AJ, editors. Bridging the Gap Trends in the Ostracode Biological & Ceelogcal Sciences. The Paleontological Society Papers. Vol. 9. 2003. pp. 203-222

[42] Forester RM. Determination of the dissolved anion composition of ancient lakes from fossil ostracodes. Geology. 1986;**14**:746-798

[43] Ito E, DeDeckker P, Eggins SM. Ostracodes and their sheel chemistry: Implications for paleohydrologic & paleoclimatic logic applications. Paleontological Society Papers. 2003;**9**:119-151

[44] Forester RM, Smith AJ. Microfossils as indicators of paleohydrology and paleoclimate. In: Paleohydrological Methods and Their Applications, Proceedings of NEA Workshop. 9-10 Nov 1992. pp. 39-57

[45] DeMott WR. The role of competition in zooplankton, succession. In: Somer U, editor. Plankton Ecology: Succession Plankton Communities. Berlin: Springer–Verlag; 1992. p. 369

[46] Forester RM, Colman SM, Reynolds RL, Keigwin LD. Lake Michigan's late quaternary limnological and climate history from ostracode oxygen isotope, and magnetic susceptibility. Journal of Great Lakes Research. 1994;**20**(1):93-107

[47] Boomer I, Whatley R, Aladin NV. Aral sea ostracoda as environmental indicators. Lethaia. 1996;**29**:77-85

[48] Forester RM, Brouwers EM. Hydrochemical parameters governing the occurrence of estuarine and marginal estuarine ostracodes: An example from south–Central Alaska. Journal of Paleontology. 1985;**59**(2):344-369

[49] Aranki JF, Mckenzie KG, Reyment ER, Reyment RA. Marine late Miocene to early Pleistocene Ostracoda of southern Spain, with the description of vejera new genus. Bulletin of the Geological Institution of the University of Uppsala, N.S. 1992;**17**:1-20

[50] Whatley RC, Arias CF, Comas-Rengifo MJ. The use of Ostracoda to detect kenoxic events: a case history

from the Spanish Toarcian. Geobios. 1994;**27**:733-741

[51] Smith AJ, Forester RM. Estimating past precipitation and temperature from fossil ostracodes. In: Proceedings of the American Nuclear Society. 5 Th International Conference and Exposition. Las Vegas: Nevada American Society of Civil Engineers & American Nuclear Society; 1994. pp. 2545-2552

[52] Pokorny V. Introduction to Marine Micropaleontology. New York: Elsevier; 1978. pp. 109-149

[53] Krushelnitskii VA, Shmeleva MA. Ostracods: Paleontological Indicators of Freshwater Ecosystems in the Complex. Study of Aquatic Ecosystems Springer; 2019

[54] Whatley RC. Recent advances in Ostracodology. Journal of Micropalaeontology. 2000;**19**(2):141-155

[55] Baumgartner PO. Ostracoda in treatise on Invertebrate paleontology, part Q: Arthropoda 3 the geological. Society of America, Inc.; 2016

[56] Schön I, Martens K. Progress and problems in ostracod systematics. Journal of Paleontology. 2012;**86**(2):214-243

[57] Crasquin-Soleau S, Brisset E. Late Triassic ostracod fauna from southeastern France: Paleoecology and paleobiogeography in Palaeogeography, Palaeoclimatology, Palaeoecology. 2014;**414**:336-346

[58] Holmes JA. Ostracoda as palaeoenvironmental indicators. Advances in Ecological Research. 2015;**52**:1-41

[59] Külköylüoğlu O, Altındağ A. Ostracod distribution in relation to water chemistry in shallow lakes of western Anatolia (Turkey). Environmental Science and Pollution Research. 2016;**23**(7):6599-6612

[60] Horne DJ, Schön I, Smith R, Martens K. What are Ostracoda? A cladistic analysis of the extant superfamilies of the subclasses Myodocopa and Podocopa (Crustacea: Ostracoda). In: Koenemann S, Jenner RA, editors. Crustacea and Arthropod Relationships, Crustacean. Vol. 16. 2005. pp. 249-273

[61] Schön I, Martens K, Van Dijk P. Lost Sex. The Evolutionary Biology of Parthenogenesis. 2009:1-615

[62] Martens BK, Savatenalinton S. An Introduction to the Study of Freshwater Ostracods. Vol. 2012. Pensoft Publishers; 2011

[63] Martens K, Schön I. Biogeography of Freshwater Ostracoda Revisited in Freshwater Animal Diversity Assessment. 2008. pp. 309-319

[64] Kaufmann RS, Martens K. Ostracoda and global change. In Palaeogeography, Palaeoclimatology, Palaeoecolog. 1997;**135**(3-4):127-141